Advances in Spatial Science

The Regional Science Series

This series contains scientific studies focusing on spatial phenomena, utilising theoretical frameworks, analytical methods, and empirical procedures specifically designed for spatial analysis. Advances in Spatial Science brings together innovative spatial research utilising concepts, perspectives, and methods relevant to both basic science and policy making. The aim is to present advances in spatial science to an informed readership in universities, research organisations, and policy-making institutions throughout the world. The type of material considered for publication in the series includes: Monographs of theoretical and applied research in spatial science; state-of-the-art volumes in areas of basic research; reports of innovative theories and methods in spatial science; tightly edited reports from specially organised research seminars.

The series and the volumes published in it are indexed by Scopus.

For further information on the series and to submit a proposal for consideration, please contact Johannes Glaeser (Senior Editor Economics) Johannes.glaeser@springer.com.

Luca Salvati • Ioannis Konaxis • Marco Maialetti

Landscape Economics

Territorial Challenges and Social Dynamics

Springer

Luca Salvati
Department of Methods and Models for Economics, Territory and Finance (MEMOTEF), Faculty of Economics
Sapienza University of Rome
Rome, Italy

Ioannis Konaxis
Department of Tourism Studies
University of Piraeus
Piraeus, Greece

Marco Maialetti
Independent researcher and lawyer
Rome, Italy

ISSN 1430-9602 ISSN 2197-9375 (electronic)
Advances in Spatial Science
ISBN 978-3-032-25262-3 ISBN 978-3-032-25263-0 (eBook)
https://doi.org/10.1007/978-3-032-25263-0

This Springer imprint is published by the registered company Springer Nature Switzerland AG
The registered company address is: Gewerbestrasse 11, 6330 Cham, Switzerland

Competing Interests The authors have no competing interests to declare that are relevant to the content of this manuscript.

Contents

Chapter 1
Local Development, Landscape Economics, and Complex Territorial Dynamics

Abstract Local development is one of the most nuanced argument when interpreting and debating how territories confront contemporary economic, social, and environmental transformations. It refers to the gradual and negotiated process through which places reshape their resource base, strengthen their institutional capacities, and cultivate collective forms of agency. What distinguishes local development from traditional economic growth models is its insistence on integrating tangible and intangible assets into a unified developmental perspective. Infrastructure, service provision, and spatial organization matter, but so too do trust, cooperation, civic engagement, and the relational networks that underpin a community's ability to act cohesively. This introductory chapter explores the internal functioning of local systems by examining the mechanisms that allow territories and landscapes to adapt, innovate, and orient themselves toward long-term transformation. Elements such as urban resilience, diversification of economic activities, and the stewardship of historically shaped landscapes, are regarded as essential drivers of territorial vitality. These dimensions illustrate how development emerges from the interplay between inherited resources, institutional learning, and the capacity of local actors to negotiate uncertainties and opportunities in a rapidly shifting global context.

Keywords Regional disparities · Indicators · Sustainable development · Agency · Local communities

Introduction

Local development can be described as a dynamic process through which a territory or city renews, enhances, and adapts its resource base in response to social, economic, and environmental transformations (Binswanger-Mkhize et al., 2010). Rather than merely representing a particular mode of economic intervention, the concept provides a comprehensive lens for understanding how places evolve within increasingly complex global systems (Malecki, 2007). It encompasses dimensions that extend far beyond traditional economic indicators—integrating social

L. Salvati et al., *Landscape Economics*, Advances in Spatial Science,
https://doi.org/10.1007/978-3-032-25263-0_1

relationships, cultural practices, political arrangements, and environmental conditions into a unified interpretive framework (Krueger et al., 2018). Because of its wide application across disciplines and contexts, the term often gathers diverse and sometimes divergent meanings, reflecting a multifaceted body of theory and practice (Cox & Mair, 1988).

In recent decades, local development has gained relevance as both a response and an alternative to the distortions produced by globalization, such as environmental degradation, territorial imbalances, and growing socioeconomic disparities (Coffey & Polese, 1985). Unlike development models grounded primarily in economic performance, it emphasizes the strengthening of territorial assets that contribute to long-term well-being (e.g., Alaimo et al., 2022). These assets include not only infrastructure and physical capital but also intangible resources, such as networks of cooperation, collective learning processes, and forms of social capital that enable communities to mobilize their internal capacities (Van der Heide & Heijman, 2013). Through such resources, cities and regions are better positioned to identify opportunities arising from global economic shifts and to transform them into drivers of local progress (Cox, 1995).

Conceived in this way, local development involves more than the activation of specific projects or policies. It represents a shift in perspective, namely an analytical and strategic "leap forward" that allows local actors to recognize the limits of their territories, to understand which dynamics arise naturally, and to identify areas where targeted interventions may generate added value (e.g., Benassi et al., 2020). Development processes rooted in territorial specificities—typically labor-intensive, less dependent on external capital flows, and closely tied to local knowledge systems—are of particular significance (Ramsay, 1996). Many of these sectors, notably those connected to green and circular economies, derive their strength from their embeddedness in place-based resources and from their relative resistance to relocation (Loiseau et al., 2016).

This perspective aligns with a systemic understanding of territory. A local system may be defined as a constellation of actors—public authorities, firms, associations, and citizens—who, under certain conditions, behave as a coordinated entity (Simon & Ando, 1961). Such systems possess their own identity and are capable of interacting with external environments while maintaining internal coherence. They are structured through dense networks of relationships that connect individuals, organizations, and institutions within a specific geographical space, and link them to broader territorial and global dynamics (Torre & Traversac, 2011). Interpreting these systems requires attention to their inherited territorial "capital," as well as to their organizational structures and their capacity for self-reproduction and innovation (Bevir, 2010).

Despite its analytical strength, this theoretical model raises several critical issues. Emphasizing local actors carries the risk of overlooking multiscalar contexts—regional, national, and global—that shape local opportunities (Barbati et al., 2013). It also assumes that local stakeholders possess the competencies necessary to understand, evaluate, and mobilize all forms of territorial resources (Smalskys et al., 2020). Moreover, translating collective (territorial) visions into effective action is often constrained by institutional fragmentation, limited cooperation, and the challenge of aligning local initiatives with broader urban, regional, and environmental policies (e.g., Biasi et al., 2017). Nevertheless, the value of local development lies precisely in its capacity to frame development as the outcome of coordinated interaction (Ceccarelli et al., 2014). It involves public, private, and hybrid actors converging around a shared vision—explicit or implicit—of how to valorize both material and immaterial resources within territories (Maillat, 1998). Local actors frequently demonstrate a superior ability to interpret and guide place-specific socioeconomic processes, making their involvement essential for producing durable and context-sensitive development pathways (Bowen, 2010).

The intrinsic competitiveness of a territory, even within global markets, depends on its ability to articulate strategies that mobilize its unique configurations of resources, knowledge, and identities—the so-called *local milieu*. In a global context with climate warming, economic recession, and deepening social inequalities, this approach offers a relevant foundation for imagining forms of progress that are both sustainable and inclusive (Jakob, 2022). Understanding local development as a process of continuous renewal and resource valorization—rather than a linear pursuit of economic growth—opens space for exploring how economic, social, cultural, political, and environmental dimensions interact to produce new opportunities (Wu et al., 2020).

This introductory chapter situates local development as a multidimensional framework through which complexities, potentials, and constraints of cities, landscapes, and territories can be understood (Chen et al., 2020). It highlights the centrality of intangible resources—such as relational networks, trust, and social capital (Acri et al., 2024)—alongside physical infrastructures (Shiell et al., 2020), and underscores how their interplay shapes the capacity of local systems, and the associated landscapes, to adapt, innovate, and orient themselves toward sustainable futures.

Landscape as a Driver of Local Development

Local development is inherently connected to the territory and its landscapes, particularly in internal or peripheral areas that have historically been marginalized by industrial development and largely overlooked in national economic planning (Domon, 2011). These internal areas, despite showing limited signs of revitalization, retain a significant (strategic) potential for sustainable territorial development. From a positive perspective, they encompass unique ecological, cultural, and social resources (Carlucci et al., 2019). From a normative standpoint, they require targeted economic, political, and regulatory interventions to unlock this potential (Sakapaji et al., 2024). A central asset of internal areas is their distinctive landscape, shaped over centuries through the interaction of natural processes and human activity (Canfora et al., 2017). This landscape not only constitutes a repository of historical memory and local identity, but also provides the foundation for innovative or alternative tourism, and place-based economic strategies (Yakymchuk et al., 2021).

Normatively, these territories should be leveraged to diversify regional economies, offering experiences and services that complement (or surpass) traditional international destinations, whose appeal has diminished (Caneva et al., 2015). Furthermore, internal areas often provide reproducible environmental resources—such as water, renewable energy, agricultural reserves, and ecosystem services—that support urban and regional metabolism, underpinning both ecological and economic functions (Kisman & Tasar, 2014). At the same time, internal areas face significant environmental and infrastructural challenges. Hydrogeological instability, exacerbated by climate change, poses substantial risks to local populations, infrastructure, and ecosystems (Merz et al., 2011).

As a representative case for advanced economies, European governance focused on urban areas and industrial districts, initially considering internal regions as marginal locations in any development process (Kløve et al., 2011). In recent decades, however, positive steps have been taken to address environmental and landscape concerns, including the introduction of landscape strategies, river basin plans, and protected areas, namely national and regional parks (Cutaia, 2016), as well as Strategic Environmental Assessment procedures (Therivel, 2012). The uncoordinated application of these planning instruments has sometimes resulted in administrative conflicts and procedural complexity (Brunori et al., 2018). Normatively, there is a clear need to integrate and streamline these tools to ensure that internal areas are managed coherently and sustainably. Aside from the Rural Development Program—which primarily provides financial support (Van Ploeg et al., 2017)—internal areas have largely remained marginalized in national economic strategies.

Given their resource endowments, the development of internal areas cannot rely on sector-specific interventions alone. Positive analysis indicates that an effective development requires integrated, multidimensional, and intersectoral policies, with the territory itself intended as an active agent. Normatively, local development should be conceived as a participatory and strategic process, in which heritage, ecological resources, social capital, and institutional capacities converge to shape context-specific development trajectories (Ansari et al., 2012). The effectiveness of such policies is proportional to the diversity of territorial resources, the engagement and coordination of local actors, and their capacity for collective action and self-organization (e.g., Biasi et al., 2019). In this framework, local development policies complement broader economic strategies by reconnecting market-oriented initiatives with the production of local collective goods and by fostering cooperation among multiple stakeholders (Kline & Moretti, 2014).

The European Union (EU) plays a crucial role in defining the institutional and regulatory framework for local development. Positive evidence shows that European structural funds have contributed to reducing regional disparities and supporting less-developed territories (Musyck & Reid, 2007; Murzyn, 2020; Muñoz-Fernández et al., 2023). These interventions should continue to target internal areas with tailored strategies that enhance territorial cohesion, strengthen local capacities, and integrate environmental, social, and economic objectives (Cini & Borragán, 2022).

The principle of subsidiarity, enshrined in the Maastricht Treaty dated 1991, emphasizes that EU action should be justified only where objectives can be better achieved at the European level, while decisions should remain as close as possible to local citizens (Von Beume, 2001; Maas, 2014; Nicolosi & Mustert, 2020). Accordingly, regions and local authorities bear primary responsibility for implementing development policies, encompassing labor market interventions, sustainable urbanization, social inclusion, cultural and tourism development, and support for innovative entrepreneurship (Grzeszczak, 2015).

In conclusion, internal areas exemplify challenges and opportunities of territorially grounded development. Positively, they possess unique ecological, cultural, and social resources that can drive sustainable growth (Becagli et al., 2013). Their potential can be realized through integrated, multidimensional policies, participatory governance, and coordinated interventions that align local action with national and European frameworks (Bajocco et al., 2018). By positioning the territory as an active subject in development, internal areas can contribute not only to economic growth, but also to the generation of social, cultural, and environmental value—fostering resilience, inclusivity, and long-term sustainability.

The Structure of This Book

Chapter 2 investigates the notion of "territory" from a multidisciplinary and "systemic" perspective, framing it not as a (merely) physical or geographical entity, but as a complex socioeconomic and cultural construct. "Territorialization" —a process that encompasses practices of naming, materialization, institutionalization, and spatial structuring of forms and functions—is conceptualized as the dynamic outcome of the interaction among material, symbolic, and relational dimensions (Kärrholm, 2007). The concept of space is explored in its multiple dimensions (namely, geometric, philosophical, economic, geographical, and cultural), illustrating how different epistemological interpretations influence both individual and collective spatial representations. Within this framework, economic space is interpreted as a fundamental determinant of regional development (Seifollahi-Aghmiuni et al., 2022).

The spatial localization of activities, scale economies, and the externalities generated by physical (and thematic) proximity, collectively contribute to the creation of competitive advantages, as well as to the reproduction of territorial disparities. Focusing on economic spaces, this contribution finally highlights the historical evolution of cartographical representations, and emphasizes the analytical significance of spatial scale in landscape interpretation (e.g., Bajocco et al., 2011). Conversely, the inherent discussion on cultural and urban spaces underscores the interdependence between social practices, settlement morphologies, and territorial transformations, particularly in relation with continuous processes of urbanization, metropolitan expansion, and sprawl (Kantor, 2019).

Rural space, by contrast, is conceptualized as a multifunctional "milieu" in which economic productivity, ecological balance, and social cohesion intersect and mutually reinforce one another. The chapter further investigates the dialectical relationship between globalization and local networks, illustrating how historical processes of industrialization and technological change have generated growth poles and core–periphery dynamics that underpin the current triadic structure of the global economy (Fischer, 2015). The notion of region is thus reinterpreted through historical, functional, and cultural lenses (sensu Burillo et al., 2020), emphasizing its emergent role as an agent of innovation, competitiveness, and governance within the knowledge-based economy (Asheim, 2012). Finally, particular attention is devoted to the interrelation between territory and environment, with specific reference to the issues of land consumption, sustainability, and environmental assessment through indicators such as the ecological footprint and carrying capacity (Francaviglia et al., 2019). The contribution concludes by stressing the necessity of integrated spatial planning and multilevel governance strategies aimed at reconciling economic development objectives with environmental preservation and social equity (Hariram et al., 2023).

Territorial Processes and Landscape Transformations

In this line of thinking, Chap. 3 illustrates the evolving concept of "landscape" as a multidisciplinary construct situated at the intersection of geography, ecology, culture, and human agency (Antrop & Van Eetvelde, 2017). From its early aesthetic and pictorial interpretations to contemporary ecological, territorial, and systemic perspectives, the landscape is no longer conceived as a visual (or static) entity, but as a dynamic configuration of interactions among natural, biotic, and anthropogenic components. Particular attention is devoted to the historical evolution of this notion, tracing its trajectory from its origins in artistic and cultural representation to its institutional codification within the European Landscape Convention (2000), which redefined landscape as both a tangible and intangible heritage—a core element of identity, social cohesion, and collective well-being (Egidi et al., 2020).

The discussion extends to cultural and anthropogenic landscapes, emphasizing the transformative role of human activity and the integration of ecological, cultural, and perceptual values within territorial systems. The analysis further addresses the pressures of urbanization, sprawl, and globalization pulses—processes that have profoundly reshaped the material and symbolic dimensions of contemporary landscapes. Finally, the work examines the conservation and valorization of cultural and environmental heritage, underscoring the landscape's dual function as a testimony of historical processes and a reservoir of biodiversity, as well as its strategic role as a foundation for sustainable (local) development. Landscapes thus emerge as heterogeneous, relational, and evolving entities, whose interpretation (and management) demand a holistic, transdisciplinary approach capable of connecting ecological processes, cultural identities, and socioeconomic transformations within an integrated framework of spatial analysis and policy action.

Chapter 4 discusses, from historical, geographical, and economic perspectives, the (apparent and latent) relationships between landscape and local development, emphasizing their interdisciplinary complexity and their centrality to future sustainability. Following a theoretical discussion that distinguishes the concept of "landscape" from that of "environment," and interpreting the former as a polysemic and dynamic construct, this contribution demonstrates how landscapes function simultaneously as a "resource"—namely a marker of cultural identity—and a driver of socioeconomic transformation (Stobbelaar & Pedroli, 2011; Förster et al., 2012; Kirchhoff et al., 2013). With this perspective in mind, the analysis investigated the main domains of interaction, including:

1. Agriculture, which has evolved from traditional systems into agro-industrial models typical of advanced economies and, more recently, into highly technological models, with profound implications for biodiversity, land consumption, and technological dependence at large (Clapp & Ruder, 2020);
2. Urbanization, which since the nineteenth century has radically reconfigured landscapes by concentrating population and economic activity, while generating social, environmental, and health imbalances (Leon, 2008), and;
3. Tourism, assumed as a strategic sector of the global economy capable of enhancing territorial value while being vulnerable to systemic crises such as the recent COVID-19 pandemic (Pashkus et al., 2021).

Further sections address (a) culture, understood both as the foundation of collective identity, and as a development factor, with landscapes serving as a vehicle of memory, belonging, and social cohesion; (b) climate change, which introduces unprecedented global challenges and requires integrated mitigation and adaptation strategies; (c) biodiversity, considered both as a common good and an economic asset whose conservation involves complex cost–benefit and ethical dimensions; and, finally, (d) social well-being, intrinsically associated with landscape perception and to conditions of equity, inclusion, and efficiency in the shared management of resources (Jakob, 2022). Finally, this contribution underscores the urgency of multiscalar governance rooted in sustainability, equity, and efficiency, referencing key policy frameworks such as the European Landscape Convention (Olwig, 2019). In such thinking, landscapes should no longer be conceived as the passive background of human activity, but as a strategic and dynamic interface integrating economy, society, culture, and nature—a fundamental dimension in steering the just transition toward sustainable and resilient models of local development (Dickeson, 2010).

The "Anthropocene" Momentum

A profound and pervasive influence of human activities on natural and socioecological processes characterizes the current geological epoch known as the Anthropocene (Reyers et al., 2018). The sudden intensification of key phenomena such as global warming, biodiversity loss, soil degradation, air and water pollution, deforestation, and stratospheric ozone depletion, demonstrates that humanity has effectively become a geological force, capable of irreversibly altering the equilibrium of ecosystems (Petrov et al., 2023). Within this framework, the discussion developed in Chap. 5 extends to the concept of sustainable development, conceived not merely as a balance among economic growth, environmental protection, and social equity, but as a comprehensive paradigm capable of integrating economic, cultural, and technological dimensions. This expanded interpretation underscores the necessity of viewing sustainability as a systemic process rather than a static equilibrium, demanding continuous adaptation and innovation. The analysis also addresses the inherent complexity and non-linearity of interactions between socio-ecological systems, landscapes, and economic structures, highlighting the decisive role of externalities and the monetary valuation of ecosystem services (Sachs, 2015).

Such relationships illustrate that environmental degradation and resource depletion are not only ecological issues but also structural failures of market and policy systems unable to internalize the true costs of environmental impacts (Dean & McMullen, 2007). The perspective adopted in this work emphasizes the urgent need for policies and strategies grounded in intergenerational responsibility and the transition toward innovative paradigms, including circular economy (Park et al., 2013). These paradigms represent essential pathways for achieving a sustainable future with technological progress, cultural evolution, and economic development well harmonized with ecological integrity and social well-being (Pike et al., 2016).

Chapter 6 examines the multifaceted relationship between human societies, economic development, and the environment, emphasizing how anthropogenic pressures have redefined ecological dynamics and landscape transformations (Sallustio et al., 2018). Situated within a broader discourse on the Anthropocene, the discussion highlights the extent to which industrialization, urbanization, and intensive resource exploitation have produced unprecedented environmental consequences, including climate change, biodiversity loss, and the widespread degradation of soil and water systems (Mukherjee et al., 2019). The text critically traces the conceptual evolution of sustainable development, from its foundational articulation in the Brundtland Report to contemporary frameworks that integrate environmental, economic, and social dimensions within a coherent and systemic perspective (Burns & Witoszek, 2012).

Particular attention is devoted to the distinction between weak and strong sustainability, and to the central role of social justice and intergenerational equity as prerequisites for achieving long-term socioecological resilience (Shemshad et al., 2025). The chapter further explores the valuation of ecosystem services and underscores the importance of advanced environmental accounting tools capable of quantifying externalities and guiding organizations toward more sustainable production and management practices (Palaniswamy & Krishnan, 2012). By linking complexity theory with the study of socioecological systems, it reveals the inherently nonlinear and interdependent nature of human–environment interactions, stressing the necessity of integrated policies and educational strategies that foster sustainability at both local and global levels (Peters, 2015). Such a transformation represents a fundamental step toward establishing a more balanced, regenerative, and enduring relationship between society, the economy, and the environment (Poister, 2010).

Local Development, Landscape Management, and Territorial Governance

By debating on the sustainable management of landscapes and natural resources, the dynamic interplay between human activity and natural processes, reflects centuries of agricultural adaptation and rural development shaping contemporary landscapes. By emphasizing their ecological, cultural, and economic significance (e.g., Ferrara et al., 2017), our discussion focuses on the fundamental roles of soil and water as essential environmental resources, the multifunctionality of agroforestry ecosystems, and the need for integrated spatial planning to prevent land degradation and hydrogeological instability (Awazi et al., 2024). Particular attention is devoted to the interdependence among landscape systems, agricultural practices, and environmental governance, demonstrating how sustainable management can simultaneously enhance ecological balance and economic resilience (Cumming, 2011). The book finally explores the digital transformation of landscape analysis through advanced technologies such as Geographic Information Systems (GIS), digital photogrammetry, and remote sensing, which enable quantitative assessment, spatial modeling, and evidence-based policymaking (Matthews et al., 2006). Ultimately, the work argues how recognizing the economic value of ecosystem services and integrating it into public policy and spatial planning are essential to achieving long-term sustainability, social well-being, and the conservation of natural capital (Stern et al., 2020). A crucial aspect of this academic debate is the governance architecture through which local development is enabled—or constrained (Budzich-Szukala, 2013). Traditional planning practices, designed around sectoral dimensions and administrative boundaries, struggle to capture the complexity of interdependent social, economic, and ecological processes. In response, integrated governance approaches have gained prominence, emphasizing policy coordination across sectors, shared decision-making, and multiscale alignment (Turok, 2012).

Such approaches seek to prevent territories from falling into the "localism trap," a condition where local actors pursue inward-looking strategies that limit their ability to engage productively with broader regional, national, or international frameworks (Ruddock & Ruddock, 2022). Acknowledging the importance of integrated governance is only a first step. The persistent gap between conceptual frameworks and practical implementation demonstrates the need for a cultural shift in how development is envisioned and enacted. Strengthening collaborative planning tools, fostering durable partnerships between public institutions and private actors, and expanding meaningful forms of citizen participation are all essential for translating local development principles into effective action (e.g., Salvati et al., 2017). These practices not only broaden the range of actors involved in shaping territorial futures, but also help anchor development strategies in the lived experiences and priorities of communities (Ledwith, 2011).

Viewed through this expanded lens, local development moves beyond the narrow pursuit of economic efficiency (Sabet & Khaksar, 2024). It becomes a comprehensive project oriented toward improving the quality of life, enhancing social cohesion, sustaining environmental integrity and safe landscapes, and building the institutional resilience necessary to navigate future challenges (Zhikharevich et al., 2021). In this sense, local development offers not just a set of policy instruments, but a framework for reimagining territorial progress—one rooted in the unique identities of places, and in the capacity of communities to act collectively and strategically over time (Vale, 2014).

Concluding Remarks

What distinguishes local development from traditional economic growth models is its insistence on integrating tangible and intangible assets into a unified developmental perspective. Landscapes, infrastructure, service provision, and spatial organization matter, but so too do trust, cooperation, civic engagement, and the relational networks that underpin a community's ability to act cohesively. A comprehensive exploration of the internal functioning of local systems (and the related landscapes) requires examining the mechanisms that allow territories and landscapes to adapt, innovate, and orient themselves toward long-term transformation. Elements such as urban–rural resilience, diversification of economic activities—and the stewardship of historically shaped landscapes—are considered not as peripheral concerns, but as essential drivers of territorial vitality. These dimensions illustrate how development emerges from the interplay between inherited resources, institutional learning, and the capacity of local actors to negotiate uncertainties and opportunities in a rapidly shifting global context.

References

Acri, M., Greffe, M. X., & Dobričić, S. (2024). Intangible matters: Cultural heritage as a driver for circular economy. In *Adaptive reuse of cultural heritage: Circular business, financial and governance models* (pp. 71–95). Springer.

Alaimo, L. S., Ciommi, M., Vardopoulos, I., Nosova, B., & Salvati, L. (2022). The medium-term impact of the COVID-19 pandemic on population dynamics: The case of Italy. *Sustainability, 14*(21), 13995.

Ansari, S., Munir, K., & Gregg, T. (2012). Impact at the 'bottom of the pyramid': The role of social capital in capability development and community empowerment. *Journal of Management Studies, 49*(4), 813–842.

Antrop, M., & Van Eetvelde, V. (2017). Landscape dynamics and evolution. In *Landscape perspectives: The holistic nature of landscape* (pp. 141–176). Springer.

Asheim, B. (2012). The changing role of learning regions in the globalizing knowledge economy: A theoretical re-examination. *Regional Studies, 46*(8), 993–1004.

Awazi, N. P., Njamnjubo, N. A., & Ambebe, T. F. (2024). Agroforestry for ecosystem services: Assessing the role of agroforestry for water regulation. *Forest.*

Bajocco, S., Ricotta, C., & Salvati, L. (2011). Land degradation versus fire: A spiral process? *Progress in Physical Geography, 35*(1), 3–18.

Bajocco, S., Smiraglia, D., Scaglione, M., Raparelli, E., & Salvati, L. (2018). Exploring the role of land degradation on agricultural land use change dynamics. *Science of the Total Environment, 636*, 1373–1381.

Barbati, A., Corona, P., Salvati, L., & Gasparella, L. (2013). Natural forest expansion into suburban countryside: Gained ground for a green infrastructure? *Urban Forestry & Urban Greening, 12*(1), 36–43.

Becagli, C., Puletti, N., Chiavetta, U., Cantiani, P., Salvati, L., & Fabbio, G. (2013). Early impact of alternative thinning approaches on structure diversity and complexity at stand level in two beech forests in Italy. *Annals of Silvicultural Research, 37*(1), 55–63.

Benassi, F., Cividino, S., Cudlin, P., Alhuseen, A., Lamonica, G. R., & Salvati, L. (2020). Population trends and desertification risk in a Mediterranean region, 1861–2017. *Land Use Policy, 95*, 104626.

Bevir, M. (2010). Interpreting territory and power. *Government and Opposition, 45*(3), 436–456.

Biasi, R., Brunori, E., Ferrara, C., & Salvati, L. (2017). Towards sustainable rural landscapes? A multivariate analysis of the structure of traditional tree cropping systems along a human pressure gradient in a Mediterranean region. *Agroforestry Systems, 91*(6), 1199–1217.

Biasi, R., Brunori, E., Ferrara, C., & Salvati, L. (2019). Assessing impacts of climate change on phenology and quality traits of *Vitis vinifera* L.: The contribution of local knowledge. *Plants, 8*(5), 121.

Binswanger-Mkhize, H. P., de Regt, J. P., & Spector, S. (2010). *Local and community driven development: Moving to scale in theory and practice*. World Bank Publications.

Bowen, S. (2010). Embedding local places in global spaces: Geographical indications as a territorial development strategy. *Rural Sociology, 75*(2), 209–243.

Brunori, E., Salvati, L., Antogiovanni, A., & Biasi, R. (2018). Worrying about "vertical landscapes": Terraced olive groves and ecosystem services in marginal land in Central Italy. *Sustainability, 10*(4), 1164.

Budzich-Szukala, U. (2013). A new approach to local development in programming EU funds—Experience from the past and challenges for the future. In *New paradigm in action* (pp. 116–138).

Burillo, P., Salvati, L., Matthews, S. A., & Benassi, F. (2020). Local-scale fertility variations in a low-fertility country: Evidence from Spain (2002–2017). *Canadian Studies in Population, 47*(4), 279–295.

Burns, T. R., & Witoszek, N. (2012). Brundtland report revisited: Toward a new humanist agenda. *Journal of Human Ecology, 39*(2), 155–170.

Caneva, G., Bartoli, F., Ceschin, S., Salvadori, O., Futagami, Y., & Salvati, L. (2015). Exploring ecological relationships in the biodeterioration patterns of Angkor temples (Cambodia) along a forest canopy gradient. *Journal of Cultural Heritage, 16*(5), 728–735.

Canfora, L., Salvati, L., Benedetti, A., & Francaviglia, R. (2017). Is soil microbial diversity affected by soil and groundwater salinity? Evidences from a coastal system in Central Italy. *Environmental Monitoring and Assessment, 189*(7), 319.

Carlucci, M., Zambon, I., Colantoni, A., & Salvati, L. (2019). Socioeconomic development, demographic dynamics and forest fires in Italy, 1961–2017: A time-series analysis. *Sustainability, 11*(5), 1305.

Ceccarelli, T., Bajocco, S., Salvati, L., & Perini, L. (2014). Investigating syndromes of agricultural land degradation through past trajectories and future scenarios. *Soil Science and Plant Nutrition, 60*(1), 60–70.

Chen, Y., Zhu, M., Lu, J., Zhou, Q., & Ma, W. (2020). Evaluation of ecological city and analysis of obstacle factors under the background of high-quality development: Taking cities in the Yellow River Basin as examples. *Ecological Indicators, 118*, 106771.

Cini, M., & Borragán, N. P. S. (2022). *European Union politics*. Oxford University Press.

Clapp, J., & Ruder, S. L. (2020). Precision technologies for agriculture: Digital farming, gene-edited crops, and the politics of sustainability. *Global Environmental Politics, 20*(3), 49–69.

Coffey, W. J., & Polese, M. (1985). Local development: Conceptual bases and policy implications. *Regional Studies, 19*(2), 85–93.

Cox, K. R. (1995). Globalisation, competition and the politics of local economic development. *Urban Studies, 32*(2), 213–224.

Cox, K. R., & Mair, A. (1988). Locality and community in the politics of local economic development. *Annals of the Association of American Geographers, 78*(2), 307–325.

Cumming, G. S. (2011). Spatial resilience: Integrating landscape ecology, resilience, and sustainability. *Landscape Ecology, 26*(7), 899–909.

Cutaia, F. (2016). *Strategic environmental assessment: Integrating landscape and urban planning*. Springer.

Dean, T. J., & McMullen, J. S. (2007). Toward a theory of sustainable entrepreneurship: Reducing environmental degradation through entrepreneurial action. *Journal of Business Venturing, 22*(1), 50–76.

Dickeson, R. C. (2010). *Prioritizing academic programs and services: Reallocating resources to achieve strategic balance*. Wiley.

Domon, G. (2011). Landscape as resource: Consequences, challenges and opportunities for rural development. *Landscape and Urban Planning, 100*(4), 338–340.

Egidi, G., Salvati, L., & Vinci, S. (2020). The long way to tipperary: City size and worldwide urban population trends, 1950–2030. *Sustainable Cities and Society, 60*, 102148.

Ferrara, C., Carlucci, M., Grigoriadis, E., Corona, P., & Salvati, L. (2017). A comprehensive insight into the geography of forest cover in Italy: Exploring the importance of socioeconomic local contexts. *Forest Policy and Economics, 75*, 12–22.

Fischer, A. M. (2015). The end of peripheries? On the enduring relevance of structuralism for understanding contemporary global development. *Development and Change, 46*(4), 700–732.

Förster, F., Großmann, R., Iwe, K., Kinkel, H., Larsen, A., Lungershausen, U., & Teichmann, M. (2012). What is Landscape? Towards a common concept within an interdisciplinary research environment. *eTopoi*, 169–179.

Francaviglia, R., Di Bene, C., Farina, R., Salvati, L., & Vicente-Vicente, J. L. (2019). Assessing "4 per 1000" soil organic carbon storage rates under Mediterranean climate: A comprehensive data analysis. *Mitigation and Adaptation Strategies for Global Change, 24*, 795–818.

Grzeszczak, R. (2015). The concept and practice of good governance in the European Union. *International Journal of Law and Political Sciences, 9*(2), 449–453.

Hariram, N. P., Mekha, K. B., Suganthan, V., & Sudhakar, K. (2023). Sustainalism: An integrated socio-economic-environmental model to address sustainable development and sustainability. *Sustainability, 15*(13), 10682.

Jakob, M. (2022). Globalization and climate change: State of knowledge, emerging issues, and policy implications. *Wiley Interdisciplinary Reviews: Climate Change, 13*(4), e771.

Kantor, P. (2019). *The dependent city revisited: The political economy of urban development and social policy*. Routledge.

Kärrholm, M. (2007). The materiality of territorial production: A conceptual discussion of territoriality, materiality, and the everyday life of public space. *Space and Culture, 10*(4), 437–453.

Kirchhoff, T., Trepl, L., & Vicenzotti, V. (2013). What is landscape ecology? An analysis and evaluation of six different conceptions. *Landscape Research, 38*(1), 33–51.

Kisman, Z. A., & Tasar, I. (2014). The key elements of local development. *Procedia Economics and Finance, 15*, 1689–1696.

Kline, P., & Moretti, E. (2014). People, places, and public policy: Some simple welfare economics of local economic development programs. *Annual Review of Economics, 6*(1), 629–662.

Kløve, B., Allan, A., Bertrand, G., Druzynska, E., Ertürk, A., Goldscheider, N., & Schipper, P. (2011). Groundwater dependent ecosystems. Part II. Ecosystem services and management in Europe under risk of climate change and land use intensification. *Environmental Science & Policy, 14*(7), 782–793.

Krueger, R., Schulz, C., & Gibbs, D. C. (2018). Institutionalizing alternative economic spaces? An interpretivist perspective on diverse economies. *Progress in Human Geography, 42*(4), 569–589.

Ledwith, M. (2011). *Community development: A critical approach.* Policy Press.

Leon, D. A. (2008). Cities, urbanization and health. *International Journal of Epidemiology, 37*(1), 4–8.

Loiseau, E., Saikku, L., Antikainen, R., Droste, N., Hansjürgens, B., Pitkänen, K., & Thomsen, M. (2016). Green economy and related concepts: An overview. *Journal of Cleaner Production, 139*, 361–371.

Maas, W. (2014). European Union citizenship in retrospect and prospect. In *Routledge handbook of global citizenship studies* (pp. 409–417). Routledge.

Maillat, D. (1998). Interactions between urban systems and localized productive systems: An approach to endogenous regional development in terms of innovative milieu. *European Planning Studies, 6*(2), 117–129.

Malecki, E. J. (2007). Cities and regions competing in the global economy: Knowledge and local development policies. *Environment and Planning C: Government and Policy, 25*(5), 638–654.

Matthews, N. A., Noble, T. A., & Breithaupt, B. H. (2006). The application of photogrammetry, remote sensing and geographic information systems (GIS) to fossil resource management. *Bulletin New Mexico Museum of Natural History and Science, 34*, 119–131.

Merz, R., Parajka, J., & Blöschl, G. (2011). Time stability of catchment model parameters: Implications for climate impact analyses. *Water Resources Research, 47*(2).

Mukherjee, S., Patel, A. K., & Kumar, M. (2019). Water scarcity and land degradation nexus in the anthropocene: Reformations for advanced water management as per the sustainable development goals. In *Emerging issues in the water environment during Anthropocene: A south east Asian perspective* (pp. 317–336). Springer Nature.

Muñoz-Fernández, G. A., Bertucci, A., Ramos-Ruiz, J. E., & Garo, M. L. (2023). The dynamics of fund absorption: Evaluating the efficacy of EU structural funds in mitigating regional inequalities—Calabrian case. *Economies, 12*(1), 10.

Murzyn, D. (2020). Smart growth in less developed regions–the role of EU structural funds on the example of Poland. *Innovation: The European Journal of Social Science Research, 33*(1), 96–113.

Musyck, B., & Reid, A. (2007). Innovation and regional development, do European structural funds make a difference? *European Planning Studies, 15*(7), 961–983.

Nicolosi, S. F., & Mustert, L. (2020). The European Committee of the Regions as a watchdog of the principle of subsidiarity. *Maastricht Journal of European and Comparative Law, 27*(3), 284–301.

Olwig, K. R. (2019). The practice of landscape 'conventions' and the just landscape: The case of the European landscape convention. In *Justice, power and the political landscape* (pp. 197–212). Routledge.

Palaniswamy, N., & Krishnan, N. (2012). Local politics, political institutions, and public resource allocation. *Economic Development and Cultural Change, 60*(3), 449–473.

Park, M., Kim, Y., Lee, H. S., Han, S., Hwang, S., & Choi, M. J. (2013). Modeling the dynamics of urban development project: Focusing on self-sufficient city development. *Mathematical and Computer Modelling, 57*(9–10), 2082–2093.

Pashkus, V., Pashkus, N., & Asadulaev, A. (2021). Impact of Covid-19 on the global tourism industry and ways to ensure high competitiveness of the territory in the global tourism market after the pandemic. In *SHS web of conferences* (Vol. 92, p. 01041). EDP Sciences.

Peters, B. G. (2015). *Pursuing horizontal management: The politics of public sector coordination.* University Press of Kansas.

Petrov, M., Nikolaeva, Z., & Dimitrov, A. (2023). The impact of anthropogenic activity on the global environment. *Science Business Society, 8*(2), 59–64.

Pike, A., Rodríguez-Pose, A., & Tomaney, J. (2016). *Local and regional development*. Routledge.

Poister, T. H. (2010). The future of strategic planning in the public sector: Linking strategic management and performance. *Public Administration Review, 70*, S246–S254.

Ramsay, M. (1996). *Community, culture, and economic development: The social roots of local action*. Suny Press.

Reyers, B., Folke, C., Moore, M. L., Biggs, R., & Galaz, V. (2018). Social-ecological systems insights for navigating the dynamics of the Anthropocene. *Annual Review of Environment and Resources, 43*(1), 267–289.

Ruddock, L., & Ruddock, S. (2022). Investment in infrastructure as a key to sustainable economic recovery: The role of the building industry. *International Journal of Strategic Property Management, 26*(6), 439–449.

Sabet, N. S., & Khaksar, S. (2024). The performance of local government, social capital and participation of villagers in sustainable rural development. *The Social Science Journal, 61*(1), 1–29.

Sachs, J. D. (2015). Investing in social capital. In *World happiness report* (pp. 152–166).

Sakapaji, S. C., García Molinos, J., Parilova, V., Gavrilyeva, T., & Yakovleva, N. (2024). Navigating legal and regulatory frameworks to achieve the resilience and sustainability of indigenous socioecological systems. *Resources, 13*(4), 56.

Sallustio, L., Pettenella, D., Merlini, P., Romano, R., Salvati, L., Marchetti, M., & Corona, P. (2018). Assessing the economic marginality of agricultural lands in Italy to support land use planning. *Land Use Policy, 76*, 526–534.

Salvati, L., Ranalli, F., Carlucci, M., Ippolito, A., Ferrara, A., & Corona, P. (2017). Forest and the city: A multivariate analysis of peri-urban forest land cover patterns in 283 European metropolitan areas. *Ecological Indicators, 73*, 369–377.

Seifollahi-Aghmiuni, S., Kalantari, Z., Egidi, G., Gaburova, L., & Salvati, L. (2022). Urbanisation-driven land degradation and socioeconomic challenges in peri-urban areas: Insights from southern Europe. *Ambio, 51*(6), 1446–1458.

Shemshad, M., Synowiec, A., Kopyra, M., & Benedek, Z. (2025). The community-driven ecosystem resilience and equity framework: A novel approach for social resilience in ecosystem services. *Sustainability, 17*(8), 3452.

Shiell, A., Hawe, P., & Kavanagh, S. (2020). Evidence suggests a need to rethink social capital and social capital interventions. *Social Science & Medicine, 257*, 111930.

Simon, H. A., & Ando, A. (1961). Aggregation of variables in dynamic systems. *Econometrica, 29*(1), 111–138.

Smalskys, V., Gavkalova, N., Babenko, K., & Zolenko, A. (2020). Efficiency of the stakeholder interaction in the context of ensuring sustainable territorial development. *Problems and Perspectives in Management, 18*(2), 340–349.

Stern, N. H., Unsworth, S., Valero, A., Zenghelis, D., Rydge, J., & Robins, N. (2020). *Strategy, investment and policy for a strong and sustainable recovery: An action plan*. Centre for Economic Performance, London School of Economics and Political Science.

Stobbelaar, D. J., & Pedroli, B. (2011). Perspectives on landscape identity: A conceptual challenge. *Landscape Research, 36*(3), 321–339.

Therivel, R. (2012). *Strategic environmental assessment in action*. Routledge.

Torre, A., & Traversac, J. B. (2011). *Territorial governance: Local development, rural areas and agrofood systems*. Springer.

Turok, I. (2012). Cities, regions and competitiveness. In *Regional competitiveness* (pp. 79–93). Routledge.

Vale, L. J. (2014). The politics of resilient cities: Whose resilience and whose city? *Building Research & Information, 42*(2), 191–201.

Van der Heide, C. M., & Heijman, W. (2013). *The economic value of landscapes* (Vol. 26). Routledge.

Van Ploeg, J. D., Renting, H., Brunori, G., Knickei, K., Mannion, J., Marsden, T., & Ventura, F. (2017). Rural development: From practices and policies towards theory. In *The rural* (pp. 201–218). Routledge.

Von Beume, K. (2001). Citizenship and the European Union. In *European Citizenship between National Legacies and Postnational Projects* (pp. 61–85). Oxford Academic.

Wu, H., Li, Y., Hao, Y., Ren, S., & Zhang, P. (2020). Environmental decentralization, local government competition, and regional green development: Evidence from China. *Science of the Total Environment, 708*, 135085.

Yakymchuk, A., Popadynets, N., Valyukh, A., Skrypko, T., & Levkov, K. (2021). Rural 'green' tourism as a driver of local economy development in the process of decentralization of power. *Agricultural and Resource Economics: International Scientific E-Journal, 7*(1), 232–259.

Zhikharevich, B. S., Klimanov, V. V., & Maracha, V. G. (2021). Resilience of a territory: Concept, measurement, governance. *Regional Research of Russia, 11*, 1–8.

Chapter 2
Space, Regions, and Landscapes: Toward a "Territorial" Perspective?

Abstract The present chapter examines the notion of territory from a multidisciplinary and "systemic" perspective, framing it not as a merely physical or geographical entity, but as a complex socioeconomic and cultural construct. "Territorialization"—a process that encompasses practices of naming, materialization, institutionalization, and spatial structuring of forms and functions—is conceptualized as the dynamic outcome of the interaction among material, symbolic, and relational dimensions. The concept of space is explored in its multiple analytical dimensions (namely, geometric, philosophical, economic, geographical, and cultural), illustrating how different epistemological interpretations influence both individual and collective (spatial) representations. Within this framework, economic space is interpreted as a fundamental determinant of regional development: the spatial localization of production activities, the presence of scale economies, and the externalities generated by physical (and thematic) proximity collectively contribute to the creation of competitive advantages, as well as to the reproduction of territorial disparities. Focusing on economic spaces, this contribution finally highlights the historical evolution of cartographical representations and emphasizes the analytical significance of geographical scale in landscape interpretation.

Keywords Economic space · Regional development · Urban sprawl · Globalization · Land consumption

Introduction

The notion of "territory" occupies a hegemonic position within both geographical and political thought (Stojanović, 2024). Broadly defined, it encompasses all that falls under the jurisdiction of a given institutional order, a sort of "territorialized space," domesticated, economically appropriated and exploited, and structured by shared practices (such as construction), as well as by ideologies and culturally codified modes of thought and action. Territory is, in this sense, a "social artifact" rather than a naturally given entity. It basically denotes a physical space subject to

L. Salvati et al., *Landscape Economics*, Advances in Spatial Science,
https://doi.org/10.1007/978-3-032-25263-0_2

continuous and inexorable processes of "territorialization" (Raffestin, 2012). Under complex thinking, territorialization entails three interdependent actions, namely (i) naming, (ii) using, and (iii) structuring a given space (Blomley, 2019).

"Naming" transforms a physical entity into a territory by assigning a specific denomination that unambiguously identifies it. Through this action whether applied to objects, beings, or places, a world of meanings is produced, embedding human histories, narratives, images, and landscapes that may be transmitted to future generations, preserved, transformed, altered or, even destroyed. Yet, this referential architecture remains problematic, insofar as it is itself a human construction. Through naming, the environment is symbolically "objectified." The second action, namely "use" (Storey, 2020), refers to the process by which a given territory is transformed into a tangible object endowed with functional utility, for instance a dam, a bridge, highways, or dwellings. This operation implies not only a modification of the material configuration of space but also an alteration of its functions and intended purposes (e.g., Cerdà et al., 2021). In this sense, reification produces a concrete and palpable transformation of both the environment and its potentialities.

The third action of territorialization consists in place structuring through an explicit subdivision into smaller, differentiated units, each endowed with a specific and distinct meaning (Schwarz & Streule, 2024). The division of the earth's surface into compartments, regions, or zones—together with the semantic ambiguities inherent in such categorizations—constitutes a foundational practice of geographical thought. Structuring thus operates through "structures of meaning," whereby space acquires territorial significance only once it is assigned a recognized function (Clemente & Salvati, 2017). Ultimately, the "territory" is the outcome of one or more of these ongoing and overlapping processes. The transition from "space" to "territory" should therefore be understood as an exclusively human construction. A territory is not merely defined by its physical extension, while being more frequently recognized as an assemblage of visible, invisible, and symbolic dimensions (Colantoni et al., 2020). Road signs, for instance, designate the name of the land traversed, while food traditions often refer to specific, and sometimes unique, territorial contexts. Enduring (topographic) features such as waterbodies, mountains, or cities are further (intrinsic) characteristics of any territory. Such instances demonstrate that these entities cannot be immediately reduced to generic categories. Rather, they attain intrinsic specificity through historical trajectories and symbolic inscriptions that constitute them as territories.

The plurality of territories across the globe may be interpreted as the cumulative outcome of historically sedimented processes that differentiate and individualize space. Each territory emerges as a unique configuration irreducible to abstract spatial categories shaped by the interplay among material form, cultural meaning, and historical depth (e.g., Colantoni et al., 2015). Some territories are protected and valued by supranational institutions owing to their designation as cultural or natural heritage, whereas others remain neglected, contested, marginalized, or deemed economically unviable. A territory is also continuously redefined through land policies, urban (or rural) planning, and cultural practices, in which both individual and collective choices converge (Colantoni et al., 2017). Environmental concerns, modes of dwelling, and cultural practices are invariably mediated through territorial frameworks (Coluzzi et al., 2022). The individual is always spatially situated, and it is through this embeddedness that the world acquires meaning.

Space, in turn, admits multiple and complex interpretations depending on its intrinsic meaning. Derived from the Latin term *spatium*, literally meaning "being open," the term operates across diverse semantic dimensions, namely geometric, philosophical, physical, geographical, economic, or cultural. Each of these dimensions (or reading frames) assumes distinct conceptual articulations depending on its relation to human perception and experience. In antiquity, within the framework of classical geometry understood as the science of relations and measurements, the connection between space and its contents was expressed in terms of proximity, distance, and magnitude dimensions that rendered space quantitatively definable through three (well-defined and measurable) axes.

Within this scientific tradition, the prevailing conception was that of natural space, understood as the projection in nature of the three-dimensional schema codified by Euclid in the "Elements of Geometry" (third century before Christ). Nevertheless, divergent and at times conflicting interpretations of space have long characterized philosophical and scientific inquiry (Wardhaugh, 2023). Early Pythagorean treatises conceived of space as an empty and unlimited backdrop from which bodies emerge (Breidenstein Jr., 2023). In the "Timaeus" (360 before Christ), Plato reconceptualized space as an internal receptacle—something that contains elements rather than existing as an external void (Betegh, 2021). This idea was subsequently elaborated by Aristotle, who defined space ("topos") as the place of a physical body, contiguous to other bodies yet lacking substantial continuity (Protin, 2021). In his formulation, we can summarize how space is "the immobile and immediate limit of the container, insofar as it is contiguous to the content."

Beyond the lexicon of philosophy, physics, and mathematics, *space* assumes more general meanings: it may refer to the environment (or substrate) within which activities, phenomena, and social relations unfold. It can also be understood as a form of "subjective intuition," perceived through the senses (particularly vision), and organized by individuals in accordance with their lived realities. Consequently, social and economic relations exert a profound influence on the perception, organization, and symbolic ordering of space (Baktybek, 2023).

Geographical Spaces

The French geographer Jean Tricart (1962) described geographical space as "the epidermis of the Earth," conceiving it as that portion of the terrestrial surface and biosphere upon which natural and anthropogenic processes interact (Chorley, 2019). Over time, this notion has acquired increasingly sophisticated theoretical articulations, enriched by more refined analytical and methodological approaches. Within an ecological perspective, the study of geographical space entails examination of the abiotic, biotic, and anthropogenic interactions that occur not only within the defined study area but also across its adjacent and functionally interconnected regions (e.g., Cutini et al., 2015). Geographical space is thus understood as a specific domain being appropriated, organized, and transformed by human societies, and may be defined as the spatial configuration and distribution of human activities and settlements across the Earth's surface (Semple, 2023).

The analysis of a given geographical space reveals its historical trajectories, developmental conditions, and the network of interdependencies that connect it to neighboring spaces. Humans represent lived space, acquire knowledge of place, and comprehend positional and directional relationships fundamental to human survival and collective organization (Doukas et al., 2023). In order to explore, and communicate, the value (and intrinsic use) of such spaces, they requires a precise representation through various forms of spatial inscription, for example, cartography.

As early as classical Greece, the concept of the "Ecumene" emerged, designating that portion of the known world in which human beings interacted with the environment and found conditions suitable for permanent settlement and productive activity (Fleetwood, 2023). The term *Ecumene*, derived from the Greek *oikein*, meaning "to inhabit," refers to the inhabited and known portion of the Earth, and reflects, in other words, "the house where we all live" (Sakupapa, 2013). Over time, "Ecumene" acquired two meanings: a geographical one, denoting the extent of the known and inhabited world, and a philosophical one, referring to a community united by shared faith, culture, or intellectual orientation.

From a geographical standpoint, the Ecumene encompassed the territories explored by geographers, merchants, and travelers who, through their journeys, disseminated knowledge and accounts of previously unknown lands and peoples. These early descriptions combined "geographical observation" with "ethnographic detail," weaving together environmental features and cultural practices encountered along the way. One of the earliest cartographic representations of the Ecumene is attributed to Anaximander of Miletus (sixth–fifth century before Christ), a disciple of Thales, who is credited with producing one of the first world maps—an attempt to model geographical space at the dawn of scientific inquiry (Papadopoulos, 2022). His successor, Hecataeus of Miletus, further advanced this cartographic tradition by representing the Ecumene within a circular frame encompassing Europe, Asia, and Libya, all encircled by the river Ocean (Irby, 2016).

As geographical knowledge expanded, representations of the Earth grew increasingly precise, and maps became instrumental in situating known territories within coherent spatial frameworks. The great explorations of the twelfth century such as those of Marco Polo to Asia and the later transoceanic voyages to the Americas rekindled a lasting interest in geography (Friedman & Figg, 2017). This resurgence was closely tied to the rediscovery of classical treatises, notably Ptolemy's Geographical Works (100–175 after Christ), and to the growing need for accurate recording of newly visited and discovered lands (Pochmalicki, 2022).

Since its origins, cartography has confronted a fundamental epistemological and technical problem, namely Earth curving (Kainz, 2020). Given the planet's near-spherical geometry, it cannot be directly represented on a planar surface such as

Table 2.1 Examples of geographical scales and applications. Source: Authors' elaboration of recent literature

Code	Minimum unit (ha)	References
1:100.00	25.0	Continental mapping
1:50000	6.25	National cartography for civil use
1:25,000	1.56	Regional (thematic) map
1:10,000	0.25	Local and topographic cartography

paper. Mathematical modeling and the progressive refinement of *cartographic projections* have thus aimed to approximate this curvature with ever-greater accuracy.

In addition to the problem of projection, the vastness of the Earth's surface necessitates a "scaling reduction" to produce readable spatial representations. Geographical scale expresses the ratio between a measured distance on a map and the corresponding actual distance on the Earth's surface. Selecting an appropriate scale is essential for analysis and interpretation of landscapes (Cao & Lam, 2023). More specifically, to render terrestrial space interpretable, it is necessary to reduce the real dimension of the Earth through mathematically defined reduction ratios, allowing maps to display proportions consistent with measurable reality. Thus, a scale of 1:50,000 indicates that the represented territory has been reduced fifty thousand times relative to its actual size. The numerator (always "1") signifies the unit of measurement on the map, while the denominator indicates the degree of reduction (Suteanu, 2023). The minimum (cartographically meaningful) unit is conventionally set at 0.25 cm^2, corresponding to an area of 5×5 mm, as indicated in Table 2.1.

Beyond its mathematical definition, geographical scale also organizes the hierarchy of spatial elements. Vertically, it progresses from the local to the regional, national, continental, and global levels. Horizontally, it may be envisioned as a system of expanding (concentric) spheres beginning with the home and extending outward to the neighborhoods, cities, metropolitan areas, physical (or administrative) regions, and, ultimately, to the global scale once again (e.g., Egidi et al., 2021). Through this multiscalar perspective, the analysis of geographical space emerges as an interpretive synthesis of physical form, ecological processes, and human organization—a dynamic "epidermis" upon which natural systems and cultural structures continuously transform and interact.

Economic Spaces

The analysis of productive resources within the economic aggregates of a given territory requires attention to both spatial and temporal dimensions (Allais & Gobert, 2019). It also underscores the importance of incorporating spatial variables into the study of regional economies and market dynamics (Irwin, 2010). When spatial factors are explicitly formalized through schemes, laws, and models governing the productive substrate, the analysis evolves into what can be defined as the economy of territories. In this perspective, "space" becomes a fundamental determinant of the competitiveness of local production systems (Egidi et al., 2020). Early twentieth-century location theory focused on the spatial distribution of economic activities across physical space, conceived as a measurable and metrically definable construct (Esposito et al., 2018). By contrast, regional development theory investigates patterns of economic growth and income distribution within specific territorial contexts, divided into regions or geographically delimited areas that exhibit a degree of internal economic homogeneity (Grillitsch & Sotarauta, 2020).

Within regional economics, the concepts of "diversified space" and "relational space" have emerged as frameworks that integrate heterogeneous research dimensions of varying analytical complexity, offering a composite perspective that reflects structural and relational approaches together (e.g., Ferrara et al., 2012). In this sense, space functions as a generator of economic advantage, as development processes are spatially uneven and tend to concentrate in specific growth poles (Kaletnik & Lutkovska, 2020). In such contexts, the dominant enterprise that catalyzes local development is often termed the driving industry. This industry attracts the investments of other firms, thereby generating mechanisms of synergy, clustering, and localized economic homogeneity.

Consequently, space becomes a constitutive factor in the economic analysis of territorial systems. Spatial and geographical proximity among firms fosters knowledge spillovers through which productive activity in a particular sector generates positive externalities extending beyond its immediate spatial boundaries (Lehmann et al., 2025). Proximity to production hubs, transportation infrastructures, and urban hierarchies therefore emerges as a key determinant in shaping both spatial dynamics and the analytical framework of regional economics (Ferrara et al., 2017). As a matter of fact, the relationship between human societies and territory underwent profound transformations with the expansion of long-distance commercial networks (Lo et al., 2020).

In preindustrial economies, the value of territory was primarily associated with its productive capacity and its ability to satisfy local subsistence needs (Horlings, 2005). Over time, land evolved from a common-use resource into an economic asset whose value depended not only on natural fertility, but increasingly on location and accessibility (Francaviglia et al., 2017). With the onset of industrialization, production became spatially concentrated in specific areas, while other parts of economic space remained comparatively underdeveloped (Ferrara et al., 2016a, b). Economies of scale consolidated this pattern, encouraging capital investment in already industrialized regions and generating a self-reinforcing, cumulative process of territorial polarization (Li et al., 2022).

The concept of economies of scale captures precisely these dynamics. It describes the relationship between the size of productive units and the average cost of output. More specifically, economies of scale refer to the inverse relationship between the expansion of production and unit cost: as output increases, the average cost per unit declines. By contrast, returns to scale denote the relationship between proportional variations in input and the corresponding change in output. If inputs double while output more than doubles, returns are increasing; if output grows proportionally less, returns are decreasing; and if output varies in direct proportion to inputs, returns are constant (Halbac-Cotoara-Zamfir et al., 2020).

Economies of scale manifest in multiple forms. A firm embedded within a dense network of suppliers may lower unit costs by purchasing larger quantities of inputs. Similarly, the same production facilities may be used to manufacture different goods, yielding joint production efficiencies. Economies of scale may also derive from the repeated (or extended) use of fixed resources without incurring additional costs. For instance, an automobile manufacturer can amortize substantial research and development expenditures for a new car by distributing these costs across a large production volume. This process reduces the average unit cost of production and strengthens the firm's competitive position within the market. Ultimately, such mechanisms illustrate how spatial concentration, technological interdependencies, and production scale coalesce to shape the territorial organization of economic activity.

Cultural Spaces

The traces left by successive layers of social organization upon a given territory are of fundamental importance for understanding the evolution and transformation of geographical space over time. Each historical phase inscribes distinctive forms, practices, and meanings upon the landscape, generating a stratified palimpsest in which physical, social, and symbolic dimensions intersect (Imbrenda et al., 2022). Within a defined geographical area, the space occupied by a given population sharing one or more homogeneous cultural traits is typically referred to as a "cultural area." Such cultural homogeneity exerts a structuring influence on the organization of territories, while the historical processes that unfold over time, in turn, modify its spatial configuration in accordance with the evolving needs and practices of local communities. These adaptations manifest in agricultural forms (e.g., terraced cultivation, crop specialization, and distinctive land-use systems), architectural expressions (e.g., vernacular building types, networks of historical roads and paths), or other material and immaterial forms of territorial adaptation that bear witness to long-term human–environment interactions. The notion of cultural area (Carrier & Gewertz, 2020) resonates with Clark Wissler's classical thinking ("The American Indian," New York, 2nd edition, 1922), as cited in Torrans (1960):

> *We saw that the natives of the New World could be grouped according to some particular cultural aspects, which gave rise to areas of food, areas of textiles, areas of ceramics,* etc. *(…) On the other hand, if we consider all the characteristic elements together and if we focus our attention on the social or tribal units, we then arrive at sufficiently well-defined groups. This will give us cultural areas, or a classification of groups according to the different cultural aspects.*

This anthropological definition underscores the spatial expression of cultural systems—an idea equally applicable to stratified historical contexts, where distinct and territorially circumscribed cultural identities remain clearly visible (Low, 2014). Here, social, economic, and cultural relations contribute to shaping the perception, use, and transformation of space, and consequently the form and meaning of the landscape. The concept of "cultural landscape," a term introduced by the German geographer Martin Schwind and later elaborated in the Italian context by Milani (2006), emphasizes the cumulative nature of landscapes as the material outcome of successive historical transformations. According to Schwind's typology, landscapes may be analytically distinguished as follows:

- Forms produced in the present, reflecting current social, economic, and technological processes
- Forms produced in the past but still active today, persisting through functional or symbolic continuity
- Forms produced in the past but no longer active, surviving as residual or fossilized traces
- Forms produced in the past yet visible only through vestiges, decipherable through archival, archaeological, or morphological analysis

This stratigraphic conception of landscape links the study of basic forms to the temporal depth of human action. The comparative analysis of distinct cultures—each autonomous, diverse, and endowed with its own systems of knowledge, techniques, and life modes—confers substantive meaning upon inhabited environments. From this perspective, cultural geography pays attention to the constitutive dimensions of human diversity such as ethnicity, language, and religion as essential factors in the spatial organization of societies. These elements not only shape the internal cohesion and symbolic identity of cultural spaces, but also determine the patterns through which human communities perceive, appropriate, and transform the territories they inhabit.

Urban Spaces

A metropolitan area may be defined as the spatial concentration of buildings, infrastructures, and functional agglomerations constituting a complex, multifunctional settlement organized by human communities in response to their collective needs. In other words, cities exist because men have organized their activities in a spatially concentrated way. This definition underscores the centrality of spatial concentration in metropolitan systems, where the colocation of activities enables the realization of agglomeration economies, that is, the economic advantages derived from proximity and density of interactions.

Extended urban spaces generally originate from an initial agglomeration that functions as the economic and functional nucleus of the future city. From this core, urban expansion proceeds centripetally, generating concentric zones that progressively merge with peripheral settlements, suburbs, or adjacent villages. The areas located at the margins of dense agglomerations are defined as peri-urban zones—namely, transitional spaces in which urban and rural systems intersect and coexist. These zones are neither entirely urban nor completely rural. They instead represent hybrid environments characterized by mixed land use, diffuse settlement patterns, and complex social and economic interdependencies. As peri-urban areas expand demographically and structurally, they tend to incorporate neighboring municipalities or satellite towns, possibly forming polycentric systems that evolve around a dominant metropolitan core.

Functional urban areas reflect the intrinsic capacity to perform activities that satisfy both internal needs (local functions) and external demands (export functions), and constitutes a key indicator for evaluating urban performance and territorial integration. The spatial morphology of a city, together with the internal articulation of its constituent zones (residential, commercial, industrial, and infrastructural), reflects the interplay between spatial form and functional capacity. From this perspective, monofunctional cities are dominated by a single economic or social activity, whereas multifunctional cities exhibit diversified functions that interact dynamically within a unified urban fabric.

From the earliest territorial aggregations to the modern metropolis (Herold et al., 2003), urban growth has followed trajectories of expansion that have had profound (and often irreversible) effects on both natural and human environments. The post-war period, in particular, witnessed a drastic transformation of urban centers. The uncontrolled expansion of built-up areas into surrounding territories led to extensive and enduring changes in land use, settlement form, and landscape structure. The interpretation of successive stages of urbanization and industrial development commonly analyzed through demographic and spatial data has allowed scholars to identify a cycle comprising four principal phases:

(a) Urbanization (1950s–1980s); this initial phase was characterized by rapid industrial expansion (Rydin, 1998), accompanied by intensive utilization of labor and raw materials. Transport costs encouraged activities to locate near major urban centers, where essential services, production facilities, and residential areas were concentrated. This outcome accelerated urban growth and large-scale rural-to-urban migration, generating the dense industrial cities typical of mid-twentieth-century development.
(b) Suburbanization (1960s–1990s); as industrial production intensified, the growing demand for land, both for productive and residential purposes, was met in peri-urban fringes (Yin, 2001). Improved transport infrastructure facilitated the decentralization of industrial and housing functions, promoting the diffusion of settlement patterns beyond the traditional urban boundaries. Peripheral areas experienced rapid growth, while the core city continued to expand, albeit at a slower rate than in the preceding period.
(c) Counter-urbanization (1970s–2000s); this stage was marked by a relative demographic and economic decline of central districts in favor of suburban (and exurban) areas, where land and housing remained more affordable and abundant (Cepeda Pico, 2023). The suburban belt became the principal engine of

metropolitan growth, while historic cores experienced population loss, economic contraction, and in some cases, physical decay.

(d) Reurbanization (1980s to present days); this stage may reflect socioeconomic transformations that have led to the functional and symbolic recovery of historical centers. Processes of economic restructuring, coupled with shifts in cultural preferences and urban policy, have revitalized central areas, reversing previous patterns of economic decline (Prysiazhniuk et al., 2020). Urban regeneration programs, investments in heritage preservation, and the rise of the knowledge economy, have all contributed to this renewed centrality. Contemporary cities are no longer defined by fixed boundaries, but by diffuse spatial configurations extending irregularly into surrounding territories. This phenomenon, commonly referred to as "urban sprawl," denotes the uncontrolled outward expansion of urban areas accompanied by declining population densities. Sprawl entails radical transformations in land use and spatial organization, undermining environmental sustainability and posing major challenges for urban governance. It represents one of the defining processes of twenty-first-century urbanization, affecting an ever-growing number of metropolitan regions worldwide.

Rural Spaces

A definition of rural space may begin with the identification of complex settlements, infrastructures, and productive activities located within the countryside. Conceptually, rural space contrasts urban space, being structured around specific forms of land use, settlement morphology, and social organization. Over time, however, a distinct dichotomy has emerged between "agricultural spaces" centered primarily on productive land use, and "rural spaces" which encompasses a broader set of ecological, social, and cultural dimensions. This conceptual differentiation was explicitly articulated in the European Commission's 1988 Communication entitled "The Future of the Rural World" (Baldock et al., 2001), which established the foundations of Community policy on agriculture and rural development. The document emphasized that

> *the notions of space or rural world go well beyond a simple geographical delimitation and refer to an entire economic and social fabric comprising a set of very different activities agriculture, crafts, small and medium-sized industries, commerce, and services.*

From this perspective, rural space can be defined as

> *the territory constituted by agricultural land, but also including uses beyond agriculture, particularly those connected with settlement and the activities of rural inhabitants.*

This broader interpretation situates rural areas as multifunctional territories in which productive, ecological, and social processes coexist and interact. According to the Council of Europe (Szilágyi, 2019), rural space can be characterized by a set of dominant features including:

- The predominance of agricultural activity in land occupation
- The presence of open, green areas with ecological functions
- Low population density
- Fragmented patterns of land ownership and tenure
- The existence of small communities fostering social proximity and interpersonal relations
- A relative prevalence of artisanal and small-scale productive activities that sustain local autonomy
- A natural landscape partly shaped by human intervention, often recognized as part of the world's cultural heritage
- A regional culture rooted in local traditions, customs, and identities

Rural space thus performs three fundamental functions (economic, ecological, and social). From an economic standpoint, rural territories host small and medium-sized enterprises in industry, crafts, and commerce that require ongoing support and innovation to remain competitive. From an ecological perspective, they preserve biodiversity, sustain forest and water resources, and provide essential ecosystem services, including carbon sequestration and recreational opportunities. Environmental protection in these contexts is intrinsically associated with the sustainable management of natural resources and landscape conservation. From a social standpoint, rural areas contribute to the cohesion of local communities, fostering social capital, intergenerational continuity, and the preservation of traditional knowledge. In this sense, rural space should be understood as multifunctional, encompassing economic production, environmental stewardship, and cultural vitality. This multifunctionality underscores the strategic importance of rural areas as spaces of both enterprise and heritage, central to the balance between development and conservation.

However, the intrinsic sustainability of rural systems is increasingly threatened by structural transformations. The pressures exerted by urban expansion, the decline of traditional agriculture, and the loss of essential services and infrastructure, have contributed to processes of rural marginalization, often culminating in depopulation or, in extreme cases, the abandonment of vast territories. For these reasons, the formulation of coherent and integrated policies protecting, revitalizing, and developing rural areas is imperative. Such policies should reconcile economic diversification with environmental conservation and social inclusion, ensuring that rural spaces continue to function as vital components of broader territorial systems.

Spatial Economic Analysis, from Global to Local

In contemporary scholarship, the study of geographical space increasingly gravitates around two interrelated yet distinct dimensions. They include (i) global spaces, interpreted through the lens of globalization, encompassing the circulation of vast flows of information, capital, goods, and technologies, and (ii) local spaces, understood as the domain of networks, regions, places, and cultural identities, where territorial dynamics manifest in concrete social and spatial configurations. To grasp the structure and functioning of the global network, it is necessary to retrace the historical evolution of the world economy, which reveals the spatial logic underlying its internal processes.

Up until the seventeenth century, industrial and economic development was concentrated primarily in Europe (Hobsbawm, 1960). The geographical organization of economic processes at that time was structured around Europe's internal system and its colonial possessions, integrated with non-European spaces. Initially, the center of this world economy was located in Spain, later shifting toward Northwestern Europe (Wallerstein, 1972). This early global system rested on free labor in Europe and forced labor in colonial territories, particularly in the Americas, which functioned as *peripheral spaces* oriented toward mining and agricultural production. Eastern Europe also occupied a peripheral position, while semi-peripheral zones—including Southern France, Italy, Sweden, and Brandenburg-Prussia—remained predominantly agrarian, sustained by sharecropping and other intermediate labor systems. Following the emergence of the American states and the Peace of Westphalia (1648), the first major conflicts for hegemony over the emerging world economy began to take shape (Kamen, 1968). Between the seventeenth and nineteenth centuries, the European system extended its reach on a global scale, consolidating colonial expansion while establishing commercial relationships with major external partners such as China, Persia, the Ottoman Empire, and the Indian subcontinent (Abu-Lughod, 1989).

The European model of triangular trade generated a polarized geographical structure, projecting internal European center–periphery dynamics onto the global stage through mechanisms of dominance and dependence. Colonies were reorganized to meet the demands of metropolitan economies, fueling specialization in agricultural monocultures, developing transport infrastructures, and prompting migratory and cultural flows aligned with the needs of dominant powers. With the onset of the Industrial Revolution, the global economic hub shifted toward Northern Europe, particularly the Anglo-Saxon world (Langton, 1984), while retaining hierarchical relationships with dependent regions. The Industrial Revolution and the regional geography of England marked the emergence of advanced economies that evolved according to Kondratiev's long waves (1925), later reformulated by Akaev and Sadovnichiy (2016):

- 1st wave (1770–1825): self-sufficiency within subsistence economies
- 2nd wave (1825–1880): specialization in the production of primary goods
- 3rd wave (1880–1930): transition from agrarian to industrial economies
- 4th wave (1930–1980): diversification of manufacturing activities
- 5th wave (after 1980): expansion of tertiary and quaternary sectors

The two world wars catalyzed the rise of new global powers, first in Germany and the United States, later in Japan and China, culminating in the triadic configuration of the global economy. Globalization thus represents a historically cumulative process, in which sequential cycles generate new interdependencies between regions and nations. The analysis of local networks requires particular attention to the first two Kondratiev waves (late eighteenth to mid-nineteenth century), when firms began to interact systematically with the territory, recognizing it as a critical determinant of production location. High transportation costs and limited mobility favored the concentration of economic activities in urban centers, which functioned as both markets and distribution hubs for finished products.

The availability of labor and the benefits of agglomeration economies reinforced these spatial patterns (Dudin et al., 2020). In technological and industrial sectors, resource extraction, transformation, and distribution are interdependent, even when geographically distant. The enterprises constituting these networks are connected through global production chains that generate outward-oriented circuits of exchange while simultaneously producing localized spillover effects that benefit proximate firms and regions. Consequently, the local economic network can be conceptualized as a territorially organized system structured along horizontal relations (flows of goods, information, and migration), and vertical relations, linking economic actors with the specific physical and environmental attributes of each place (e.g., climate, soils, resource endowment).

Within this framework, space may be interpreted as a relational structure composed of nodes (cities), lines (transportation and communication networks), areas (patterns of land use and settlements), flows (circulations of goods, capital, and information), and hierarchies (urban, regional, and global systems). Central cities in the world economy, together with their hierarchical interconnections with surrounding territories, underscores the non-random logic of metropolitan agglomerations. The Industrial Revolution and the subsequent expansion of factories profoundly reconfigured traditional hierarchies, producing two emblematic (spatial) outcomes, namely (i) cities with a strong productive specialization either diversified across sectors, or concentrated in a dominant industry, and (ii) industrial districts, characterized by dense networks of small and medium-sized enterprises embedded within localized territorial systems.

Territory and Regions

The term *region* derives from the Latin *regio*, meaning "direction" or "area under rule." In the "Encyclopédie" (1780), the region was defined as

> *a large expanse of land inhabited by several contiguous peoples of the same nation, possessing borders and limits, and generally subject to a king or a despot.*

This early conception reflects the political and administrative meaning of the notion—emphasizing control, boundaries, and territorial sovereignty. In the late eighteenth and nineteenth centuries, geographers began to interpret the region primarily through the lens of physical geography, identifying it with a homogeneous portion of the Earth's surface characterized by uniform natural features such as climate, relief, vegetation, and hydrography. The "natural region" thus became a key analytical category in classical geography, corresponding to the spatial articulation of environmental factors.

At the turn of the twentieth century, however, the concept underwent a profound transformation with the emergence of geographical possibilism, a theoretical vision associated primarily with Paul Vidal de la Blache and the French school of human geography. Possibilism rejected the environmental determinism that had dominated nineteenth-century geographical thought, positing instead that space should be interpreted in relation with time, culture, and human agency. Within this framework, geography was redefined as the study of the relationship between societies and their environments, where the individual and the community are viewed as active geographical agents, capable of shaping, organizing, and transforming the territory through action and decisions.

In this perspective, the region is no longer a passive, naturally delimited entity, but rather a historically and culturally constructed space—the outcome of dynamic interactions between physical conditions, social practices, and temporal processes. Subsequently, the concept of regions evolved in accordance with the principle that nature offers, intended as a possibility to humans, rather than imposing absolute constraints. This theoretical shift, central to geographical possibilism, emphasizes that environments provide a range of potential conditions that human societies can exploit through their cultural, technological, and historical development. Hence, human agency expressed through settlement patterns, productive organizations, and social structures, emerges as a fundamental geographical factor, shaping territorial space. At the European level, the subsequent debate on borderless and integrated regions—particularly in the context of the European Union's regional and cohesion policies—has centered on the criteria of homogeneity, cohesion, and territorial coherence. These principles underpin the recognition of regions as spatial frameworks for planning, governance, and socioeconomic integration. Within this conceptual evolution, several typologies of regions have been identified as:

1. Formal (or natural) regions, defined with homogeneous physical characteristics, such as climate, geomorphology, or vegetation
2. Functional regions, identified on the basis of spatial cohesion and internal interdependence, often expressed through economic flows, transport networks, or labor markets
3. Historical or real regions, grounded in shared social and cultural attributes, including language, traditions, and life modes, which confer identity and continuity to territorial organization
4. Administrative (or artificial) regions, established through political or institutional decisions, serving purposes of governance, statistical analysis, and policy implementation

This multidimensional classification reflects the progressive complexification of the notion of "region," integrating physical, functional, cultural, and institutional dimensions.

The Evolving Notion of "Regions"

A natural region constitutes a portion of geographical territory whose boundaries are clearly delimited by physical features such as mountain chains, or river systems, or by other natural elements, including climate, vegetation, fauna, and soils. These regions exhibit a high degree of physical and environmental homogeneity, which renders them easily identifiable and distinct from their surrounding areas (Vujadinović & Šabić, 2017). By contrast, horizontal relations and the cohesive interactions among their constituent parts defined functional regions. Their coherence arises not from natural uniformity, but from the density of economic, social, and infrastructural exchanges that take place within them.

The spatial extent of a functional region is therefore determined by the strength and frequency of internal flows and linkages. A paradigmatic example is the metropolitan district, which concentrates and organizes flows of people (e.g., commuters), goods, and services from surrounding municipalities, generating pronounced economic and social interdependencies. Historical or real regions are instead defined by their cultural and anthropological homogeneity. In these territories, traditions and identities are deeply rooted and continue to manifest through enduring customs, local practices, and cultural events that shape everyday life. Their persistence over time confers a strong sense of belonging and territorial identity.

A key driver of regional development and local competitiveness lies in their capacity to construct and manage networks that interact effectively with globalization processes. In this regard, the region assumes the role of an intermediary space, situated between the local and the global scales, and functioning as a locus of cooperation, exchange, and strategic alliances that promote sustainable growth and innovation. Accordingly, regional analysis extends beyond administrative, physical, or historical dimensions, encompassing emerging processes that assign a delegated and active role in fostering local development and innovation to each specific region.

Geographical inquiry, in this context, examines the spatial distribution of activities, flow circulation, product distribution, and consumption dynamics across multiple scales (local, national, continental, and global). From an economic standpoint, regions can be classified according to their organizational structure and productive capacity, often manifesting as specialized industrial districts characterized by sectoral homogeneity. A crucial element of regional competitiveness concerns the conversion of knowledge into innovation and, consequently, into economic growth. Targeted innovation policies have thus fostered the emergence of territorial models of innovation, which may be summarized as (i) endogenous innovation, developed within local scientific and research networks; (ii) creative application of knowledge to production systems; and (iii) imitative innovation, based on the adoption and adaptation of external technological solutions. Building on these premises, five typologies of innovation-oriented regions have been identified within the old continent (Capello & Lenzi, 2013):

1. European Scientific area, including German regions and Central European areas such as Vienna, Brussels, and Syddanmark (Denmark). These regions exhibit

strong (internal) innovation processes, with specialization in general-purpose technologies such as computing, transport infrastructure, energy systems, and the internet.
2. Applied Science area, comprising core metropolitan regions in Austria, Belgium, Denmark, France, Finland, Germany, Luxembourg, Ireland, and Sweden, as well as peripheral capital cities such as Prague, Athens, and Lisbon. These areas are highly innovative but tend to specialize in non-general-purpose technologies (Powell & Dusdal, 2017).
3. Smart Technology Application area, encompassing highly urbanized regions such as Northern Spain, Madrid, the Czech Republic, Northern Portugal, Italy, the French Alps, the Netherlands, Ljubljana, Sweden, and the United Kingdom. These regions demonstrated limited endogenous knowledge creation, but a strong capacity for technological application and innovative product development (Sánchez-Corcuera et al., 2019).
4. Intelligent and Creative Diversification area, including Mediterranean regions, such as Central Italy and part of Spain, selected agglomerations in Slovakia and Poland, and parts of Finland and the United Kingdom. Their competitiveness relies primarily on creativity and external knowledge acquisition.
5. Imitative Innovation area, associated with regions in Eastern and Southern Europe (e.g., Bulgaria, Hungary, Lithuania, Malta, Poland, Romania, Slovakia, Southern Italy, and Greece), which show limited (internal) innovative capacity but attract substantial foreign direct investment, thus presenting significant potential for technological upgrading.

This territorial taxonomy of innovative regions provides a robust framework for analyzing the spatial differentiation of innovation patterns and processes in Europe and has become central to the reform of European regional policy, particularly in the context of smart specialization strategies.

Social Geography and Regions

Social geography may be defined as the discipline that examines geographical space in its relationship with human societies, analyzing how individuals and communities adapt to, transform, and organize it in accordance with their needs (Semple, 2023). At the same time, it explores the ways in which space itself responds and reacts to these functional transformations. For every human need, there exists a corresponding spatial manifestation, and the evaluation of the ensuing social impacts becomes essential when formulating territorial development strategies and planning forecasts.

The relationship between society and environments thus reveals the dynamic interaction between communities and the material space they inhabit, exposing the structural logics and functional mechanisms that guide its use and organization. A predictive analysis of future interventions such as the construction of infrastructures of public utility (e.g., landfills, waste-to-energy plants, or transportation corridors) is particularly relevant, as such projects can reshape local systems, influencing both social relations and spatial morphologies.

In this perspective, human geography constitutes the branch of geography investigating the spatial organization of human activity, including distribution, location, and interaction (Grčić, 2012). Drawing on the classical concept of the Ecumene (the ensemble of inhabited portions of the Earth's surface) contemporary studies highlight the coexistence of ecumenical (inhabited) and ecumenical (uninhabited) areas, the latter still covering nearly 20% of the planet's surface. The subecumenical zone, defined as a territory inhabited only intermittently or seasonally, often in response to climatic and environmental constraints, lies between these two categories.

A key variable in the spatial analysis of population distribution is demographic density, calculated as the ratio between the number of inhabitants and the surface area they occupy. However, this variable represents only an average statistical indicator, which may obscure significant intraterritorial disparities. The example of Egypt is paradigmatic: in 2013, the country's population exceeded 83 million inhabitants, with an average density of 84 inhabitants per km^2. Yet this value masks the extreme unevenness of the demographic distribution. As Mountjoy (1952) observed, given that deserts constitute the vast majority of Egypt's land area, nearly 98% of the population is concentrated within the Nile Valley and Delta, which account for merely 3% of the total territory. Population densities thus range from 3000 inhabitants per km^2 in Cairo to as few as 1–2 inhabitants per km^2 in El Wadi region.

Based on these preliminary assumptions, two fundamental factors underpin the study of demographic dynamics, namely (i) the natural balance, that is, the difference between births and deaths (Harte, 2007), and (ii) the migratory balance, corresponding with the difference between immigration and emigration (Hassan, 2002). Together, these indicators provide an accurate measure of vitality and stability of population systems, allowing for the assessment of demographic growth, decline, or equilibrium within a given territorial context. In the broader framework of human–environment relationships, demography assumes a key role, given its profound influence on social organization and the spatial configuration of societies. Ultimately, landscapes emerge as the most evident expression of these processes. They embody the tangible synthesis of the interaction between natural features and human action, serving as a key discriminant in the identification and differentiation of regions. The combination of shared landscape traits and human practices contributes to shaping the territorial identity and cultural coherence of each geographical region.

Territory and the Environment

"Environment" as a noun derives from the Latin term *ambire*, meaning literally "to surround." In its most general sense, environments constitute a complex system of physical, chemical, and biological elements—both living and non-living—together with the web of relationships that connect them, within which all organisms on the planet coexist (Kaushika et al., 2016). In its very meaning, the "environment" can be defined as "a set of physical conditions, chemical characteristics of water, air, and soil, and living beings that surround humankind." The *Oxford English Dictionary* (1989) further describes it as "the conditions under which any person or thing lives or develops; the totality of influences that modify or determine the development of life and character."

In ecological terms, the environment encompasses "everything with which a living being comes into contact, influencing positively or negatively its life cycle." This term may be extended to the entire planet, whose unique physical characteristics enable the persistence of life. Within this framework, it is common to distinguish between biotic factors, referring to the living components of ecosystems, and abiotic factors, corresponding to non-living elements such as air, water, minerals, and geomorphological structures. From an anthropocentric perspective, the environment can be understood as the ensemble of physical, floristic, faunistic, historical, cultural, and social elements that both surround and interact with human beings. The dynamic interplay among these components gives rise to ecosystems, defined as the functional units in which biotic and abiotic elements coexist and exchange energy and matter in a state of relative equilibrium. At the planetary scale, the Earth is organized into several interconnected spheres, each encompassing a specific set of physical and biological processes (Trofimov et al., 2024):

- Biosphere: the portion of the Earth where conditions permit the existence of life, comprising the lower atmosphere, the entire hydrosphere, and the superficial layer of the lithosphere
- Lithosphere: the solid outer shell of the Earth, including the crust and the upper mantle, and incorporating the pedosphere, the superficial layer in which soils develop
- Atmosphere: the gaseous envelope that protects life on Earth, functioning both as a shield against cosmic radiation and as a regulator of climatic conditions
- Hydrosphere: the totality of the planet's water, covering approximately 70% of the Earth's surface

Within these spheres, ecosystems are often categorized according to dominant vegetation formations, giving rise to large ecological units known as biomes. Climatic variables such as temperature, humidity, and precipitation, combine with the vertical stratification of vegetation to generate environmental differentiation (Woodward et al., 2004). Despite their relative ecological homogeneity, biome boundaries are gradual and transitional rather than rigidly defined. Over the course of history, human activity has modified the environment, often producing severe degradation of natural systems. The physical degradation of the biosphere, which manifests in multiple forms, is among the most critical anthropogenic impacts. They include atmospheric pollution, resulting from greenhouse gas emissions, industrial discharges, and depletion of the ozone layer, as well as water pollution, caused by airborne contaminants absorbed by aquatic systems and by chemical substances released from industrial and agricultural processes. Soil pollution, associated with the accumulation of toxic residues and heavy metals, and latent processes of land consumption, deforestation, and unsustainable exploitation of natural resources, additionally lead to environmental degradation, soil depletion and early desertification (Seinfeld & Pandis, 2016).

In response to escalating environmental degradation, the international community initiated a process of global environmental governance aimed at promoting coordinated strategies for sustainability. A key milestone in this evolution was the United Nations Conference on Environment and Development (UNCED), held in Rio de Janeiro in June 1992 (Johnson, 1992). The Rio Conference represented a historic turning point in environmental policy and in the conceptualization of sustainable development, bringing together representatives from 172 countries including 108 heads of state and over 2400 delegates from non-governmental organizations. Among its most significant outcomes was the formulation of a global action plan that would later culminate in the Kyoto Protocol (ratified in 1997), which established binding commitments for the reduction of greenhouse gas emissions (Böhringer, 2003; Grubb, 2004; Bonneuil et al., 2021).

At Rio, strategic priorities were articulated that continue to underpin international environmental governance today, including:

(a) The inherent transformation of production and consumption patterns to reduce the release of toxic substances (e.g., leaded fuels and industrial waste)
(b) The exploration and adoption of alternative energy sources to replace fossil fuels, recognized as the principal drivers of global climate change
(c) The implementation of emission-reduction strategies in the transport sector to mitigate the health effects of smog and particulate matter
(d) The promotion of integrated solutions to address water scarcity, increasingly regarded as one of the most critical challenges of the twenty-first century

The 1992 Rio Summit thus laid the conceptual foundations for sustainable development, climate policy, and environmental governance, positioning the environment not merely as a physical entity, but as a systemic and interdependent dimension of global society (Marong, 2003).

Academic Definitions and Operational Practices

Territories can be understood both as a product of human activity and as a humanized surface of the Earth. Etymologically, the term derives from the Latin *terra* and the suffix *-torium*, indicating a portion of terrestrial space characterized by specific physical or natural features. Territory denotes a delimited portion of land, a material and spatial entity organized through political–administrative, socioeconomic, and cultural structures, and shaped by economic activities and populations (Satheesh & Joseph, 2025). In its broadest conception, the notion of territory transcends physical characteristics to encompass the unfolding network of social relations, economic processes, and cultural identities. Agriculture, industrial production, tourism, and metropolitan dynamics coexist and interact, giving rise to a complex and integrated system. Within this system, the natural environment serves as the foundational matrix, yet it is continuously transformed by human intervention and development processes (e.g., Salvati & Zitti, 2005).

Over the last few decades, some advanced economies have undergone profound socioenvironmental transformations. The transitions from premodern to modern (and, subsequently, post-modern) societies, have exerted mounting pressure on fragile ecosystems, triggering widespread processes of environmental degradation and unsustainable development (Salvati & Zitti, 2009). In response to these challenges, the United Nations General Assembly established the World Commission on Environment and Development (WCED) in 1983, with the mandate to formulate a global agenda for change (Kumar, 2020). The Commission's final report presented in 1987 under the chairmanship of Gro Harlem Brundtland, and widely known as the Brundtland Report, introduced the seminal definition of sustainable development, according to which

> *humanity has the ability to make development sustainable, ensuring that it meets the needs of the present without compromising the ability of future generations to meet their own needs* (Singh, 2024).

This formulation established the ethical and operational foundation of contemporary environmental policy, positioning sustainability as a cross-cutting principle in spatial planning and resource management (e.g., Tomao et al., 2017). Within this framework, a territory can be interpreted both as a product of anthropogenic transformation and as a homogeneous physical and functional surface. Humanity, through the appropriation and exploitation of natural resources, creates the material conditions for its own survival. However, these resources are finite and unevenly distributed, necessitating an urgent rethinking of human–environment interactions. Environmental awareness indirectly stimulates measuring, monitoring, and managing landscape degradation through scientifically grounded tools and indicators. Among these, particular relevance is attributed to the capacity for sustainable development, the ecological footprint, land consumption, and other variables that capture the extent and intensity of human impact on ecosystems. To better illustrate the relationship between human activity and the environment—particularly within the context of advanced economies—two analytical dimensions will be briefly examined, namely the ecological footprint and land consumption.

The notion of "ecological footprint" was introduced in the early 1990s by Mathis Wackernagel and William Rees at the University of British Columbia (Biemer, 2021). Originally conceived as a methodological framework to quantify humanity's demand on natural systems, it has evolved into a universal metric adopted by scientists, policymakers, enterprises, and international institutions to assess resource use and guide sustainable development strategies (Wackernagel & Galli, 2007). The ecological footprint measures the biologically productive land and water area required to supply the resources consumed by a specific population, given existing technologies, and to assimilate the waste it produces, including carbon emissions (Rees & Wackernagel, 2012). It therefore constitutes a tangible indicator of the equilibrium (or disequilibrium) between anthropogenic demand and the planet's regenerative capacity. This analytical tool enables the evaluation of sustainability thresholds and supports the development of policies aimed at balancing human development with ecological resilience.

Territory and Land Management

Soil is widely recognized as a non-renewable resource, as its natural formation occurs over geological timescales that far exceed human temporal horizons. In recent years, the interaction between humans and soil has become a central concern in debates on landscape conservation, sustainable territorial management, and ecological resilience. According with the Thematic Strategy for Soil Protection, adopted by the European Commission in 2006 (Arias-Navarro et al., 2023), soil is defined as "the upper layer of the Earth's crust, composed of mineral particles, organic matter, water, air, and living organisms, representing the interface between land, air, and water, and hosting a large part of the biosphere." This complex and dynamic system performs essential ecological and biogeochemical functions, including food production for human sustenance, the preservation of biodiversity, the regulation of nutrient and carbon cycles, and the maintenance of equilibrium within the biosphere. Consequently, soil degradation implies the loss (or impairment) of these vital functions (Lal, 2009). At the European level, multiple processes and human pressures contributing to degradation have been identified (European Commission, 2006; 2012a). The most significant include:

- Erosion, driven by the mechanical action of water and wind
- Decline in organic matter, leading to reduced fertility and the disruption of ecological balance
- Contamination, whether localized or diffuse, resulting from industrial, agricultural, or urban emissions
- Land consumption, defined as the conversion of agricultural or natural surfaces into artificial land use
- Soil sealing, that is, the permanent covering of land with impermeable materials
- Compaction, often caused by heavy machinery or overgrazing, which reduces soil porosity and biological activity
- Loss of biodiversity, associated with the destruction or fragmentation of habitats
- Salinization, originating from natural processes or unsustainable irrigation practices
- Landslides and flooding, increasingly intensified by inappropriate land use
- Desertification, representing the terminal stage of soil degradation, particularly in ecologically fragile environments

Within this framework, soil protection emerges as a strategic dimension of environmental sustainability, requiring coordinated, multiscalar governance strategies to mitigate anthropogenic pressures and safeguard the ecological, economic, and social functions of this key resource. From a territorial perspective, the economic processes that stimulate the establishment of industrial, commercial, and residential infrastructures have drawn significant scholarly attention to the phenomenon of land consumption, intended as a progressive (and often irreversible) conversion of natural or agricultural areas into artificial surfaces. The most tangible and immediate manifestation of this phenomenon lies in the expansion of sealed soils, unable to assure essential ecosystem services (Ferrara et al., 2016a, b). Such surfaces include urbanized zones, industrial platforms, paved and unpaved roads, extraction sites, landfills, construction yards, courtyards, and public squares, as well as airports, ports, railway infrastructures, sports facilities, and large-scale energy installations.

In recent years, photovoltaic fields and other renewable-energy infrastructures have also contributed to this expansion patterns, raising new debates regarding the balance between energy transition and soil conservation (Hassan et al., 2024). From a functional standpoint, all these uses share the capacity to induce soil sealing that is, the impermeabilization of the surface layer which disrupts natural ecological processes, alters the hydrological cycle, reduces infiltration and groundwater recharge, and ultimately limits the availability of land for agricultural, ecological, and cultural purposes (Mancino et al., 2016).

Territory and Economic Development

Territories have been increasingly recognized as fundamental variables when studying economic dynamics, owing to the close interdependence between production systems and social structures. The progressive political decentralization of recent decades, which has transferred various functions to local and regional authorities, has accentuated the need for a comprehensive understanding of territorial systems through socioeconomic and spatial indicators (Storonyanska et al., 2020). Among these are the identification of institutional and functional areas, the analysis of labor markets and industrial structures, the characterization of the entrepreneurial fabric, and the assessment of financial systems, infrastructure, and innovation capacity (e.g., Salvati & Carlucci, 2014).

The economic development of a given territory depends not only on its internal resources but also on the ability of urban centers to establish economic networks with the surrounding areas. These connections foster the creation of extended territorial coalitions, systems of cities, and neighboring agglomerations that operate as integrated regional entities (Imbrenda et al., 2023). In this perspective, the relationship between territories and urban environments can be analyzed through a set of quantitative and qualitative indicators. Density and compactness of urban settlements represent priority metrics in contemporary spatial planning. The growth of sealed and impermeable surfaces, which occurs predominantly in peri-urban areas, manifests through the development of new residential districts and dispersed settlements. Sprawl refers to a rapid (outward) expansion of cities and settlements, typically characterized by low-density residential development, single-use zoning, and a strong dependence on private transportation. Measuring indicators of metropolitan density and settlement compactness (Barnes et al., 2001; Ewing & Hamidi, 2015; Song et al., 2020) is therefore essential for evaluating the extent of land consumption, as well as its consequences for the inherent depletion of natural capital and the loss of soil functionality.

Empirical evidence underscores the magnitude of the phenomenon. The analysis of urban–territorial relationships should therefore consider the increasing complexity of urbanization processes and the transformation of territorial systems (Wilson et al., 2017). Effective environmental risk mitigation requires a deep understanding of key factors such as land consumption, the spread of impermeable surfaces, soil quality degradation, irregular settlement patterns, and land abandonment. Beyond its ecological repercussions, urban sprawl entails profound social implications. The proliferation of low-density residential areas fosters daily commuting predominantly by private vehicles to access workplaces and services concentrated in compact urban centers (Benassi et al., 2020). This pattern contributes to higher air pollution levels, increased energy consumption, and significant transformations. Newly developed suburban neighborhoods often lack a coherent sense of community and territorial identity, isolating their inhabitants from the traditional urban fabric, and giving rise to what have been described as 'cities within the city' spaces—functionally connected yet socially detached (Vaughan, 2007).

Accordingly, the quality of urban centers can no longer be assessed solely on the base of physical morphology or infrastructural efficiency (Lehmann, 2010). Rather, it should be evaluated through the relationship between human communities and their territorial context, integrating environmental, social, and economic dimensions (Bibri & Krogstie, 2017). Public green spaces, air quality and emissions, water and waste management, sustainable mobility systems, tourism sustainability, and social cohesion, are among the key determinants of urban quality. All of them contribute to the resilience and livability of contemporary cities.

References

Abu-Lughod, J. L. (1989). *Before European hegemony: The world system AD 1250–1350*. Oxford University Press.

Akaev, A. A., & Sadovnichiy, V. A. (2016). A closed dynamic model to describe and calculate the Kondratiev long wave of economic development. *Herald of the Russian Academy of Sciences, 86*(5), 371–383.

Allais, R., & Gobert, J. (2019). Conceptual framework for spatio-temporal analysis of territorial projects. *Environmental Impact Assessment Review, 77*, 93–104.

Arias-Navarro, C., Panagos, P., Jones, A., Amaral, M. J., Schneegans, A., Van Liedekerke, M., & Montanarella, L. (2023). Forty years of soil research funded by the European Commission: Trends and future. A systematic review of research projects. *European Journal of Soil Science, 74*(5), e13423.

Baktybek, I. (2023). Socio-economic perspectives in terms of social and physical space interactions. *Annals of Spiru Haret University Economic Series, 23*(2), 94–103.

Baldock, D., Dwyer, J., Lowe, P., Petersen, J. E., & Ward, N. (2001). *The nature of rural development: Towards a sustainable integrated rural policy in Europe*. Synthesis Report, New York.

Barnes, K. B., Morgan, J. M., III, Roberge, M. C., & Lowe, S. (2001). *Sprawl development: Its patterns, consequences, and measurement*. Towson University.

Benassi, F., Iglesias-Pascual, R., & Salvati, L. (2020). Residential segregation and social diversification: Exploring spatial settlement patterns of foreign population in southern European cities. *Habitat International, 101*, 102200.

Betegh, G. (2021). Plato on illness in the *Phaedo*, the *republic*, and the *Timaeus*. In W. Plato's Timaeus (Ed.), *Proceedings of the tenth symposium Platonicum Pragense* (pp. 228–258). Palgrave Macmillan.

Bibri, S. E., & Krogstie, J. (2017). Smart sustainable cities: A review and synthesis of relevant literature. *Sustainable Cities and Society, 31*, 183–212.

Biemer, J. R. (2021). *Our environmental handprints: Recover the land, reverse global warming, reclaim the future*. Bloomsbury.

Blomley, N. (2019). The territorialization of property in land: Space, power and practice. *Territory, Politics, Governance, 7*(2), 233–249.

Böhringer, C. (2003). The Kyoto protocol: A review and perspectives. *Oxford Review of Economic Policy, 19*(3), 451–466.

Bonneuil, C., Choquet, P. L., & Franta, B. (2021). Early warnings and emerging accountability: Total's responses to global warming, 1971–2021. *Global Environmental Change, 71*, 102386.

Breidenstein, J. I., Jr. (2023). Pythagoras and Heraclitus. In *Nietzschean, feminist, and embodied perspectives on the Presocratics: Philosophy as partnership* (pp. 141–189). Springer Nature.

Cao, C., & Lam, N. S. N. (2023). Understanding the scale and resolution effects in remote sensing and GIS. In *Scale in remote sensing and GIS* (pp. 57–72). Routledge.

Capello, R., & Lenzi, C. (2013). Territorial patterns of innovation and economic growth in European regions. *Growth and Change, 44*(2), 195–227.

Carrier, J. G., & Gewertz, D. B. (2020). *The handbook of sociocultural anthropology*. Routledge.

Cepeda Pico, P. (2023). De-urbanization: From the shock to the revolution of a new urban logic. In *Urbicide: The death of the city* (pp. 751–776). Springer Nature.

Cerdà, A., Daliakopoulos, I. N., Terol, E., Novara, A., Fatahi, Y., Moradi, E., Salvati, L., & Pulido, M. (2021). Long-term monitoring of soil bulk density and erosion rates in two *Prunus persica* (L.) plantations under flood irrigation and glyphosate herbicide treatment in La Ribera district, Spain. *Journal of Environmental Management, 282*, 111965.

Chorley, R. J. (2019). *Introduction to geographical hydrology: Spatial aspects of the interactions between water occurrence and human activity*. Routledge.

Clemente, M., & Salvati, L. (2017). "Interrupted" landscapes: Post-earthquake reconstruction in between urban renewal and social identity of local communities. *Sustainability, 9*(11), 2015.

Colantoni, A., Ferrara, C., Perini, L., & Salvati, L. (2015). Assessing trends in climate aridity and vulnerability to soil degradation in Italy. *Ecological Indicators, 48*, 599–604.

Colantoni, A., Grigoriadis, E., Sateriano, A., Sarantakou, E., & Salvati, L. (2017). Back to Von Thünen: A southern European perspective on mono-centric urban growth, economic structure and non-urban land decline. *International Planning Studies, 22*(3), 173–188.

Colantoni, A., Egidi, G., Quaranta, G., D'Alessandro, R., Vinci, S., Turco, R., & Salvati, L. (2020). Sustainable land management, wildfire risk, and the role of grazing in Mediterranean urban–rural interfaces: A regional approach from Greece. *Land, 9*(1), 21.

Coluzzi, R., Bianchini, L., Egidi, G., Cudlin, P., Imbrenda, V., Salvati, L., & Lanfredi, M. (2022). Density matters? Settlement expansion and land degradation in peri-urban and rural districts of Italy. *Environmental Impact Assessment Review, 92*, 106703.

Cutini, A., Chianucci, F., Giannini, T., Manetti, M. C., & Salvati, L. (2015). Is anticipated seed cutting an effective option to accelerate transition to high forest in European beech (*Fagus sylvatica* L.) coppice stands? *Annals of Forest Science, 72*(5), 631–640.

Doukas, Y. E., Salvati, L., & Vardopoulos, I. (2023). Unraveling the European agricultural policy sustainable development trajectory. *Land, 12*(9), 1749.

Dudin, M., Levina, E., Alferov, V., & Zasko, V. (2020). N. D. Kondratiev's long wave theory and modernity: The digital age. *Revista Inclusiones, 7*, 380–394.

Egidi, G., Halbac-Cotoara-Zamfir, R., Cividino, S., Quaranta, G., Salvati, L., & Colantoni, A. (2020). Rural in town: Traditional agriculture, population trends, and long-term urban expansion in metropolitan Rome. *Land, 9*(2), 53.

Egidi, G., Cividino, S., Paris, E., Palma, A., Salvati, L., & Cudlin, P. (2021). Assessing the impact of multiple drivers of land sensitivity to desertification in a Mediterranean country. *Environmental Impact Assessment Review, 89*, 106594.

Esposito, P., Patriarca, F., & Salvati, L. (2018). Tertiarization and land use change: The case of Italy. *Economic Modelling, 71*, 80–86.

Ewing, R., & Hamidi, S. (2015). *Measuring sprawl and its impacts*. Routledge.

Ferrara, A., Salvati, L., Sateriano, A., & Nolè, A. (2012). Performance evaluation and cost assessment of a key indicator system to monitor desertification vulnerability. *Ecological Indicators, 23*, 123–129.

Ferrara, A., Kelly, C., Wilson, G. A., Nolè, A., Mancino, G., Bajocco, S., & Salvati, L. (2016a). Shaping the role of 'fast' and 'slow' drivers of change in forest-shrubland socio-ecological systems. *Journal of Environmental Management, 169*, 155–166.

Ferrara, A., Salvati, L., Sateriano, A., Carlucci, M., Gitas, I., & Biasi, R. (2016b). Unraveling the 'stable' landscape: A multi-factor analysis of unchanged agricultural and forest land (1987–2007) in a rapidly expanding urban region. *Urban Ecosystems, 19*(2), 835–848.

Ferrara, C., Carlucci, M., Grigoriadis, E., Corona, P., & Salvati, L. (2017). A comprehensive insight into the geography of forest cover in Italy: Exploring the importance of socioeconomic local contexts. *Forest Policy and Economics, 75*, 12–22.

Fleetwood, L. (2023). Histories of habitability from the oikoumene to the Anthropocene. *Wiley Interdisciplinary Reviews: Climate Change, 14*(5), e840.

Francaviglia, R., Di Bene, C., Farina, R., & Salvati, L. (2017). Soil organic carbon sequestration and tillage systems in the Mediterranean Basin: A data mining approach. *Nutrient Cycling in Agroecosystems, 107*(1), 125–137.

Friedman, J. B., & Figg, K. M. (2017). *Routledge revivals: Trade, travel and exploration in the middle ages (2000): An encyclopedia*. Routledge.

Grčić, M. (2012). Concept of cultural geography as a science. *Glasnik—Geografsko društvo Republike Srpske, 15*, 1–22.

Grillitsch, M., & Sotarauta, M. (2020). Trinity of change agency, regional development paths and opportunity spaces. *Progress in Human Geography, 44*(4), 704–723.

Grubb, M. (2004). The economics of the Kyoto protocol. In *The economics of climate change* (pp. 92–134). Routledge.

Halbac-Cotoara-Zamfir, R., Smiraglia, D., Quaranta, G., Salvia, R., Salvati, L., & Giménez-Morera, A. (2020). Land degradation and mitigation policies in the Mediterranean region: A brief commentary. *Sustainability, 12*(20), Article 8313.

Harte, J. (2007). Human population as a dynamic factor in environmental degradation. *Population and Environment, 28*(4), 223–236.

Hassan, F. A. (2002). Population dynamics. In *Companion encyclopedia of archaeology* (pp. 672–713). Routledge.

Hassan, Q., Viktor, P., Al-Musawi, T. J., Ali, B. M., Algburi, S., Alzoubi, H. M., & Jaszczur, M. (2024). The renewable energy role in the global energy transformations. *Renewable Energy Focus, 48*, 100545.

Herold, M., Goldstein, N. C., & Clarke, K. C. (2003). The spatiotemporal form of urban growth: Measurement, analysis and modeling. *Remote Sensing of Environment, 86*(3), 286–302.

Hobsbawm, E. J. (1960). The seventeenth century in the development of capitalism. *Science & Society*, 97–112.

Horlings, E. (2005). Pre-industrial economic growth and the transition to an industrial economy. In *Early modern capitalism* (pp. 97–114). Routledge.

Imbrenda, V., Coluzzi, R., Di Stefano, V., Egidi, G., Salvati, L., Samela, C., Simoniello, T., & Lanfredi, M. (2022). Modeling spatio-temporal divergence in land vulnerability to desertification with local regressions. *Sustainability, 14*(17), 10906.

Imbrenda, V., Coluzzi, R., Mariani, F., Nosova, B., Cudlinova, E., Salvia, R., Quaranta, G., Salvati, L., & Lanfredi, M. (2023). Working in (slow) progress: Socio-environmental and economic dynamics in the forestry sector and the contribution to sustainable development in Europe. *Sustainability, 15*(13), 10271.

Irby, G. L. (2016). Greek and Roman cartography. In *A companion to science, technology, and medicine in ancient Greece and Rome* (pp. 817–835). Palgrave Macmillan.

Irwin, E. G. (2010). New directions for urban economic models of land use change: Incorporating spatial dynamics and heterogeneity. *Journal of Regional Science, 50*(1), 65–91.

Johnson, S. P. (1992). Did we really save the earth at Rio? *European Environmental Law Review, 1*, 81.

Kainz, W. (2020). Cartography and the others: Aspects of a complicated relationship. *Geo-spatial Information Science, 23*(1), 52–60.

Kaletnik, G., & Lutkovska, S. (2020). Innovative environmental strategy for sustainable development. *European Journal of Sustainable Development, 9*(2), 89–89.

Kamen, H. (1968). The economic and social consequences of the thirty years' war. *Past & Present, 39*, 44–61.

Kaushika, N. D., Reddy, K. S., & Kaushik, K. (2016). The environment. In *Sustainable energy and the environment: A clean technology approach* (pp. 17–29). Springer.

Kumar, R. (2020). The United Nations and global environmental governance. *Strategic Analysis, 44*(5), 479–489.

Lal, R. (2009). Soil degradation as a reason for inadequate human nutrition. *Food Security, 1*(1), 45–57.

Langton, J. (1984). The industrial revolution and the regional geography of England. *Transactions of the Institute of British Geographers*, 145–167.

Lehmann, S. (2010). *The principles of green urbanism: Transforming the city of sustainability*. Earthscan.

Lehmann, E. E., Schenkenhofer, J., & Vismara, S. (2025). Hidden champions and knowledge spillovers: Innovation-enhancing agglomeration effects and niche technology specificity. *Small Business Economics*, 1–21.

Li, G., Zhang, R., Feng, S., & Wang, Y. (2022). Digital finance and sustainable development: Evidence from environmental inequality in China. *Business Strategy and the Environment, 31*(7), 3574–3594.

Lo, F. Y., Rey-Martí, A., & Botella-Carrubi, D. (2020). Research methods in business: Quantitative and qualitative comparative analysis. *Journal of Business Research, 115*, 221–224.

Low, S. (2014). Spatializing culture: An engaged anthropological approach to space and place. In *The people, place, and space reader* (pp. 34–38). Routledge.

Mancino, G., Nolè, A., Salvati, L., & Ferrara, A. (2016). In-between forest expansion and cropland decline: A revised USLE model for soil erosion risk under land-use change in a Mediterranean region. *Ecological Indicators, 71*, 544–550.

Marong, A. (2003). From Rio to Johannesburg: Reflections on the role of international legal norms in sustainable development. *Georgetown International Environmental Law Review, 16*, 21.

Milani, R. (2006). *The idea of landscape and the perspectives of a common European policy* (p. 3). Negotiating European Cultural Identity.

Mountjoy, A. B. (1952). *Egypt's population problem. Transactions and Papers (Institute of British Geographers), 18*, 121–135.

Papadopoulos, A. (2022). Ancient geography: Some markers. In *Mathematical geography in the eighteenth century: Euler, Lagrange and Lambert* (pp. 9–46). Springer.

Pochmalicki, L. (2022). Marco polo in the *Novus Orbis*: The reception and circulation of the *Devisement du monde* in the renaissance. *Digital Philology: A Journal of Medieval Cultures, 11*(2), 332–350.

Powell, J. J., & Dusdal, J. (2017). Science production in Germany, France, Belgium, and Luxembourg: Comparing the contributions of research universities and institutes to STEM and health. *Minerva, 55*(4), 413–434.

Protin, C. (2021). Modern incarnations of the Aristotelian concepts of continuum and topos. *arXiv preprint* https://arxiv.org/abs/2105.05889

Prysiazhniuk, O., Plotnikova, M., Buluy, O., & Yakobchuk, V. (2020). Reurbanization as a solution of socio-environmental and economic problems. *Management Theory and Studies for Rural Business and Infrastructure Development, 42*(1), 41–50.

Raffestin, C. (2012). Space, territory, and territoriality. *Environment and Planning D: Society and Space, 30*(1), 121–141.

Rees, W., & Wackernagel, M. (2012). Urban ecological footprints: Why cities cannot be sustainable—And why they are a key to sustainability. In *The urban sociology reader* (pp. 157–165). Routledge.

Rydin, Y. (1998). Postwar planning 1950s–1970s. In *Urban and environmental planning in the UK* (pp. 28–50). Macmillan Education.

Sakupapa, T. C. (2013). Local ecumenism in the Zambian context. *South African Perspectives on Notions and Forms of Ecumenicity, 2*, 156.

Salvati, L., & Carlucci, M. (2014). A composite index of sustainable development at the local scale: Italy as a case study. *Ecological Indicators, 43*, 162–171.

Salvati, L., & Zitti, M. (2005). Land degradation in the Mediterranean basin: Linking bio-physical and economic factors into an ecological perspective. *Biota, 6*, 67–77.

Salvati, L., & Zitti, M. (2009). Substitutability and weighting of ecological and economic indicators: Exploring the importance of various components of a synthetic index. *Ecological Economics, 68*(4), 1093–1099.

Sánchez-Corcuera, R., Nuñez-Marcos, A., Sesma-Solance, J., Bilbao-Jayo, A., Mulero, R., Zulaika, U., & Almeida, A. (2019). Smart cities survey: Technologies, application domains and challenges for the cities of the future. *International Journal of Distributed Sensor Networks, 15*(6), 1550147719853984.

Satheesh, K., & Joseph, C. (2025). *Spatial planning in India: A preliminary classification of existing models* (doctoral dissertation), Politecnico di Torino.

Schwarz, A., & Streule, M. (2024). Territorial subjectivities: The missing link between political subjectivity and territorialization. *Progress in Human Geography, 48*(3), 275–291.

Seinfeld, J. H., & Pandis, S. N. (2016). *Atmospheric chemistry and physics: From air pollution to climate change*. Wiley.

Semple, E. C. (2023). *Influences of geographic environment: On the basis of Ratzel's system of anthropo-geography*. Good Press.

Singh, V. (2024). Environment, development, and sustainability. In *Textbook of environment and ecology* (pp. 327–339). Springer Nature.

Song, S., Liu, Z., He, C., & Lu, W. (2020). Evaluating the effects of urban expansion on natural habitat quality by coupling localized shared socioeconomic pathways and the land use scenario dynamics–urban model. *Ecological Indicators, 112*, 106071.

Stojanović, Đ. (2024). Space, territory and sovereignty: Critical analysis of concepts. *Human Territorial Behaviors and Power Politics in Assaf Gavron's The Hilltop: A Spatial Re-Examination, 89*(1), 111–184.

Storey, D. (2020). Territory and territoriality: Retrospect and prospect. In *A research agenda for territory and territoriality* (pp. 1–24). Edward Elgar.

Storonyanska, I., Patytska, K., Hrynchyshyn, I., & Chemerys, V. (2020). Spatial disproportions in development of territorial community under conditions of administrative and financial decentralization. *Agricultural and Resource Economics: International Scientific E-Journal, 6*(4), 43–62.

Suteanu, C. (2023). Scale as ratio in space. In *Scale: Understanding the environment* (pp. 93–123). Springer.

Szilágyi, J. E. (2019). The changing concept of rural community and its importance in connection with the transfer of agricultural land. *Zbornik Radova, 53*, 633.

Tomao, A., Quatrini, V., Corona, P., Ferrara, A., Lafortezza, R., & Salvati, L. (2017). Resilient landscapes in Mediterranean urban areas: Understanding factors influencing forest trends. *Environmental Research, 156*, 1–9.

Torrans, T. (1960). General works on the American Indian: A descriptive bibliography. *Arizona and the West, 2*(1), 79–103.

Trofimov, V. T., Korolev, V. A., & Kharkina, M. A. (2024). Ecological functions as fundamental integral characteristics of peculiarities of abiotic environments of ecosystems. *Moscow University Geology Bulletin, 79*(3), 467–476.

Vaughan, L. (2007). The spatial syntax of urban segregation. *Progress in Planning, 67*(3), 205–294.

Vujadinović, S., & Šabić, D. (2017). The importance of regions in geographical research. *Zbornik Radova—Geografski fakultet Univerziteta u Beogradu, 65*(1a), 195–208.

Wackernagel, M., & Galli, A. (2007). An overview on ecological footprint and sustainable development: A chat with Mathis Wackernagel. *International Journal of Ecodynamics, 2*(1), 1–9.

Wallerstein, I. (1972). Three paths of national development in sixteenth-century Europe. *Studies in Comparative International Development, 7*(2), 1–12.

Wardhaugh, B. (2023). *Encounters with Euclid: How an ancient Greek geometry text shaped the world*. Princeton University Press.

Wilson, G. A., Kelly, C. L., Briassoulis, H., Ferrara, A., Quaranta, G., Salvia, R., Detsis, V., Curfs, M., Cerda, A., El-Aich, A., Liu, H., Kosmas, C., Alados, C. L., Imeson, A., Landgrebe-Trinkunaite, R., Salvati, L., Naumann, S., Danwen, H., Iosifides, T., Kizos, T., Mancino, G., Nolè, A., Jiang, M., & Zhang, P. (2017). Social memory and the resilience of communities affected by land degradation. *Land Degradation & Development, 28*(2), 383–400.

Woodward, F. I., et al. (2004). Vegetation–climate boundaries: The role of vertical stratification. *Global Change Biology, 10*(4), 496–506.

Yin, J. S. (2001). *The community development system: Urban politics and the practice of neighborhood redevelopment in two American cities from the 1960s to the 1990s* (doctoral dissertation). Cornell University, New York.

Chapter 3
Envisioning the "Territorial" Perspective of Landscapes

Abstract The present chapter explores the evolving concept of landscape as a multidisciplinary construct situated at the intersection of geography, ecology, culture, and human agency. From its early aesthetic and pictorial interpretations to contemporary ecological, territorial, and systemic perspectives, landscapes are no longer conceived as a visual or static entity but as a dynamic configuration of interactions among natural, biotic, and anthropogenic components. Particular attention is devoted to the historical evolution of landscapes, tracing specific trajectories up to the institutional codification within the European Landscape Convention (2000), which redefined landscape as both a tangible and intangible heritage—a core element of identity, social cohesion, and collective well-being. The discussion extends to cultural and anthropogenic landscapes, emphasizing the transformative role of human activity and the integration of ecological and cultural values within territorial systems. The analysis further addresses the intrinsic pressure of urbanization, industrialization, and globalization processes that have profoundly reshaped the material and symbolic dimensions of contemporary landscapes. Finally, the chapter examines the conservation and valorization of cultural and environmental heritage, underscoring the landscape's dual function as a testimony of historical processes and a reservoir of biodiversity, as well as its strategic role as a foundation for sustainable territorial development. In conclusion, "landscape" emerges as a heterogeneous, relational, and evolving entity, whose interpretation and management demand a holistic, transdisciplinary approach capable of connecting ecological processes, cultural identity, and socioeconomic transformations within an integrated framework of spatial analysis and policy action.

Keywords Cultural heritage · Landscape Convention · Local systems · Urbanization · Sustainability

L. Salvati et al., *Landscape Economics*, Advances in Spatial Science,
https://doi.org/10.1007/978-3-032-25263-0_3

Introduction

The evolution of the operational definition of "landscape" is evident when focusing on some statements made by various authors who have defined landscape as:

> *the character of a region of the Earth in its totality (*Von Humboldt—early nineteenth century)
>
> *the overall appearance of a country, as it moves our aesthetic feelings* (Porena, 1892)
>
> *a set of natural beauties connected to literature, art or history* (Ricci, 1905)
>
> *an abstract synthesis of visible elements* (Biasutti, 1963)
>
> *a stretch of land surface that can be encompassed by the gaze from a given point of view* (Sestini, 1963)
>
> *the face of the earth, a visible and sensitive face, which summarizes and expresses the overlapping of multiple phenomena* (Barbieri, 1991)
>
> *It is not only something to build or protect, but [...] something to recognize, perceive, listen to and describe. [...] The landscape is the 'hypostasis' of history in the territory. What has been in ethics, in aesthetics, in architecture, in philosophy, in progress or decadence, in famine or abundance, in war or peace, in history or myth, in moments of intense religiosity or agnosticism, is written in the landscape profile and all interpretable when culture, like a demiurge, intervenes and helps with enlightenment* (Andreotti, 1996)
>
> *the means to come into contact in the most immediate and direct way with a place, with a society; it becomes an intermediary, an indicator, an autobiography. The landscape is the face of a country, of a society* (Pinchemel-Pinchemel, 1996)
>
> *A theatre in which man is both actor and spectator* (Turri, 1974–1998)
>
> *the particular physiognomy of a territory determined by its physical, anthropic, biological and ethnic characteristics* (Wikipedia, 2017)
>
> *a part of the territory that can be seen from a specific point. The term is used in particular with reference to characteristic panoramas for their natural beauty, or to places of historical and artistic interest, but also, more generally, to the entire complex of natural assets that are a fundamental part of the ecological environment to be defended and conserved* (Treccani Encyclopedia, 2017).

From the chronological evolution of these definitions, it becomes evident that human perception of landscapes, as well as the degree of cultural participation in shaping its outward form, has progressively evolved in parallel with the growing inclusion of biotic and abiotic components at the core of its conceptual framework. This semantic shift has progressively aligned the notion of landscape with an ecological and systemic perspective, moving beyond earlier territorial or purely descriptive geographical interpretations. In this regard, Zonneveld (1979) defines the landscape as

> *a portion of the Earth's surface composed of a complex system resulting from the interaction of rocks, water, air, plants, animals, and humans, which through its physiognomy constitutes a recognizable entity.*

Similarly, Forman and Godron (1986) describe it as

a heterogeneous territory consisting of a cluster of interacting ecosystems that recur across space in a similar configuration.

By integrating these perspectives within a multidisciplinary and ecological framework, the landscape can be defined as

a portion of the Earth's surface characterized by a homogeneous structural organization and a distinctive spatial configuration of physical, biotic, and anthropogenic components, associated with equally well-defined functional processes.

In this sense, the landscape may be conceived as an ecological mosaic, structured through clusters, patches, and corridors within a foundational matrix, reflecting the interdependence of form, functions, and processes in the organization of terrestrial space.

The Historical Evolution of Landscapes

Landscapes represent the outcome of a long-standing interaction between human communities and natural environments, evolving in parallel with the development of social, economic, and cultural practices that have progressively shaped countries over the centuries. The origins of the term *landscape* have been the subject of extensive scholarly inquiry. Among the most illuminating interpretations is that proposed by the French semiologist Jeanne Martinet in "Le paysage: signifiant et signifié" (Talento, 2024). According to Martinet, the concept of "landscape" first emerged in the Dutch language during the second half of the fifteenth century with the term *landschap* (Talento et al., 2019). This term was adopted by Flemish painters to denote a shift of focus from representation as a mere artistic exercise toward the autonomous aesthetic value of the represented subject. In their works, nature itself becomes the central protagonist, inaugurating a radically new visual and conceptual perspective. As recalled by the painter Joachim Patenier, cited by Martinet, this innovation

reversed the scale of values recognized up to that moment, immeasurably enlarging the landscape at the expense of the figures, now relegated to a subordinate role.

While confirming the Dutch origin of the word—albeit in its corrected form "lantscap"—Yves Luginbühl, an agricultural engineer, geographer, and one of the principal supporter of the European Landscape Convention, offers an alternative interpretation (Luginbühl & Howard, 2016). He traces the term to the compound "lant" (territory) and "scap," corresponding to the German *schaft* (community). It is, therefore, a term that inherently links territory and community, originating in the Dutch, Danish, and German coastal regions inhabited by populations settled in marginal, marshy lands along the North Sea. These territories, rendered habitable through the construction of *terpen* embankments built by accumulating soil to defend against tides were managed autonomously, outside feudal authority. In his book entitled "Dimensions du paysage," published by the Council of Europe,

Luginbühl further argues that *lantscap* emerged within the context of customary land management systems, known in English as "commons," namely shared territories collectively governed and exploited by local communities (Luginbühl, 2007).

Over subsequent centuries, the notion evolved depending on the use of each European idiom, while maintaining a sort of common dynamics: the gradual attenuation of the original meaning, rooted in the relationship between communities and their territories, and the simultaneous accentuation of the visual and representational dimension (Van Putten et al., 2020). This process redefined the very foundation of "landscape," shifting its emphasis toward the interplay between reality and image, substance and form, representation and the represented (Cosgrove, 2008). In this context, Michael Jakob, in his book entitled "Le paysage," highlights a series of paradoxes intrinsic to both the word and the concept (Jakob, 2006). According to Jakob, landscape

> *is neither measurable, identifiable, nor an objective construct, but rather a phenomenon that resists any attempt at stabilization; its representation, whether verbal or visual, collides with the fluctuating, open, and at times unsettling identity of the phenomenon itself.*

The specific term *landscape* is thus used

> *both for the representation such as a painting or work of art and for the very reality that presents itself to someone as a landscape: the represented world.*

By contrast, Annalisa Calcagno Maniglio, in a contribution entitled "For a Quality Landscape. Dialogue on Failures and Delays in the Implementation of the European Landscape Convention" adopts a more ontological and scientific perspective. She argues that landscape

> *must be understood as a physical entity composed of a set of interrelated natural, biotic, and abiotic elements, within which the signs and traces of innumerable stratifications are imprinted characterized by a network of actions and relationships that have bound and continue to bind humans to their territory. As an existing reality, the landscape can be studied through scientific methods that enable its objective analysis and the comprehension of its structures, processes, and evolutionary mechanisms.*

(Recent) Landscape Visions

Observing landscapes is an ontological process inseparable from a more operational process of acquiring knowledge into multiple interrelated factors shaping landscapes. In contemporary Europe, the visual dimension of landscapes no longer corresponds with an immediate (or static) viewpoint but rather emerges as the summary of successive perceptions that unfold across time and space. Yet the landscape cannot be reduced to a "visual" dimension alone: it also encompasses olfactory, acoustic, and tactile dimensions (e.g., Imbrenda et al., 2021). In this sense, the "mental" representation of landscapes constitutes a vast multisensory spectrum, in which territorial elements activate (associative and interpretative) processes in multiple directions (Pallasmaa, 2012).

Douglas Porteous, in "Landscapes of the Mind" (1990, cited in McLean, 2021), explored this expanded conception of perception by introducing a taxonomy of experiential categories: the "smellscape" or olfactory landscape, the "soundscape" or acoustic landscape, the "bodyscape" or corporeal landscape, the "inscape" as interior landscape, the "homescape" or domestic landscape, the "childscape" or landscape of childhood, the "deathscape" or landscape of death and, finally, the more elusive "otherscapes." Each term, by virtue of the semantic richness of English lexicon, immediately evokes a distinctive experiential image, broadening the interpretative horizons of landscape studies. Landscape perception is further conditioned by temporal and situational variables such as the time of a day, the season, the angle of light, and atmospheric conditions, all of which influence visual and emotional responses. Equally significant are the physical and psychological characteristics of the observer: age, gender, health status, emotional state, and social context (Egidi et al., 2021).

Consequently, landscape perception is inherently dynamic and variable; the same place may appear profoundly different across moments of observation or states of mind. To systematize these mechanisms, scholars have identified three primary typologies of landscape perception: the Neutral-based Landscape (NbL), the Perceived Landscape, and the Cultural Landscape, each shaped by the observer's cognitive and cultural background (Johnson et al., 2020). The perceived landscape thus emerges as the composite result of these interdependent perceptual dimensions (Parra-Sanchez et al., 2023).

In recent decades, landscape studies have increasingly intersected with biodiversity and ecological research, promoting integrated approaches that combine ecological, cultural, and perceptual components, which identifies physiographic units and homogeneous territorial systems through specific cartographic analysis (Kreplin et al., 2021). By estimating ecological value and environmental fragility, empirical methodologies allow defining landscape units that jointly consider natural and cultural features. This ecological orientation has deep roots in the European scientific tradition (Forman, 2014). As early as the 1930s, Carl Troll defined landscape as the "integrated study of the environment, in which vertical and horizontal relationships between ecosystems are analyzed," emphasizing both intraecosystem processes and interecosystem interactions (Van Eetvelde & Aagaard Christensen, 2023). Troll's perspective grounded the early works of the European School of Landscape Ecology, which soon highlighted the normative and applied dimensions of landscape science (Troll, 2007). Within this framework, functional indicators are fundamental tools for ecological planning, structured around the recognition of observation scales and recurrent spatial patterns (Kairis et al., 2013).

Building on these principles, Forman and Gordon (1995) systematized landscape ecology through the Patch–Corridor–Matrix Model (Bozkurt & Basaraner, 2024), defining the landscape as "a mosaic of interacting ecosystems, which repeats itself with a recognizable spatial configuration across extensive areas." Recognizing spatial heterogeneity at the level of populations, communities, and ecosystems further underscores the decisive role of human activities in shaping landscapes (e.g., Incerti et al., 2007). The European Landscape Convention (2000) consolidated these developments by redefining the landscape as a central component of individual and collective well-being, qualifying rights and responsibilities concerning its care, management, and planning (Selman & Swanwick, 2010). This normative framework reaffirms the dual nature of landscapes, as a (i) material system of ecological relationships and as a (ii) cultural construct mediated by human perception and action.

The European Landscape Convention

The European Landscape Convention (ELC) was signed in Florence on 20 October 2000 (Jones et al., 2007). Promoted by the Council of Europe, it constitutes the first international treaty devoted exclusively to the protection, management, and planning of landscapes, while promoting international cooperation in the field of landscape policy. Unlike previous instruments focused primarily on exceptional or scenic landscapes, ELC applies to all landscapes—outstanding and degraded alike—extending its scope to the entire territory of member states (Dempsey & Wilbrand, 2017). The Convention's preamble and four chapters, comprising 18 articles, formalize the recognition of the landscape as an essential component of people's surroundings, an expression of cultural and natural diversity, and a foundation of individual and collective identity (Phillips, 2014). Its overarching objective is to serve as an instrument of social, political, and legal innovation, ensuring that European citizens enjoy a framework of rights and responsibilities with regard to the quality of their living environment. Chapter I (Article 1) provides a conceptual framework that clarifies Convention's key definitions, including:

- Landscape, intended as a specific part of a given territory, as perceived by populations, whose character results from the action and interaction of natural and human factors
- Landscape policy, regarded as the formulation of principles, strategies, and guidelines for safeguarding, managing, and planning landscapes
- Landscape quality targets, namely the aspiration of populations to specific characteristics of their living environment, as formulated by public authorities
- Landscape protection, intended as the conservation and maintenance actions aimed at preserving significant (or characteristic) aspects of a given landscape, justified by its heritage value
- Landscape management, defined as actions designed, from a sustainable development perspective, to guide and harmonize transformations driven by social, economic, and environmental change
- Landscape planning, qualified as forward-looking measures intended to enhance, restore, or create landscapes.

Article 2 defines the scope of application, which encompasses all types of landscapes (natural, rural, urban), and includes terrestrial, inland water, and marine environments. Article 3 specifies the objectives of the Convention, namely to promote landscape protection, management, and planning, and to organize European cooperation in this domain. Among the general commitments established in Article 5, each contracting party undertakes to:

(a) Recognize the landscape in its domestic law
(b) Establish and implement coherent landscape policies
(c) Ensure public participation, in collaboration with local and regional authorities, in defining and applying such policies
(d) Integrate landscape considerations into land-use, cultural, environmental, agricultural, social, and economic policies

Article 6 introduces specific measures organized into five key areas:

1. Awareness-raising—to strengthen public, institutional, and professional understanding of landscape values and roles
2. Training and education—to promote (a) specialist training for practitioners, (b) multidisciplinary programs for professionals, and (c) educational courses in schools and universities
3. Identification and assessment—each contracting party is required to identify and analyze its landscapes, monitor their evolution, and assess them in light of values expressed by local populations
4. Landscape quality objectives—to be defined for each landscape type through participatory processes
5. Implementation—to operationalize the instruments for safeguarding, managing, and planning landscapes

The Convention further underscores international cooperation among contracting states and the active involvement of local and regional authorities in identifying, evaluating, and managing landscapes, reinforcing subsidiarity and democratic participation. The Convention's most significant contribution to national legislation lies in its departure from a predominantly aesthetic conception of landscape, advancing instead a systemic and multidimensional understanding (Olwig, 2019). The landscape is thus recognized not only for its physical and natural attributes but also for the cultural values intrinsic in the local communities (Lanfredi et al., 2023). Consequently, landscape planning should address not only outstanding (e.g., protected) areas but also "ordinary" and "everyday" landscapes, acknowledging their centrality in daily life and their role as expression of collective identity and cultural continuity (Strecker, 2024).

Cultural Landscapes and Social Transformations

The notion of "landscape" has long been a central object of investigation across multiple disciplines—geography, ecology, architecture, anthropology, and the humanities. Each discipline interprets landscape within own theoretical frameworks and the adopted spatial–temporal scales (Antrop & Van Eetvelde, 2017).

Consequently, a variety of perspectives have emerged. A broadly shared interdisciplinary understanding conceives the landscape not as a mere aggregation of discrete elements, but as a systemic entity in which each component acquires meaning through its reciprocal relationship with the others (Lanfredi et al., 2022). These interrelations determine the structural and functional coherence of the system, endowing each landscape with a distinctive and recognizable identity expressed through specific physical, biotic, and anthropogenic characteristics (Mancino et al., 2016). Within the domain of ecology, landscape is understood as a dynamic unit composed of interacting ecosystems, whose configuration and evolution are shaped by the combined influence of natural processes and human actions (Hersperger et al., 2021).

Thanks to the continuous exchange among these elements—whether ecological events or cultural interventions modulates both the structure and function of landscapes over time—anthropogenic and natural/biotic dimensions are both indispensable to landscape analysis and definition (Marchi et al., 2018). The study of landscape thus requires a holistic approach capable of integrating ecological dynamics with cultural, social, and historical processes. In this sense, "landscape" etymological definitions go beyond the mere natural environment composed of biotic and abiotic components (e.g., Mariani et al., 2018). They instead encompass the territory as the stage upon which historical, cultural, and social processes unfold, generating the material and symbolic configurations that shape collective identity (Jackson, 2022), which fuses "land" referring to soil or terrain with the Germanic verb *schaffen*, meaning "to shape" or, better, "to give form." This etymology highlights the active role of human agency in transforming and structuring the environment.

Through the analysis and identification of landscape constituent elements—especially those responsible for determining its morphology and distinctive characteristics—it becomes possible to grasp the essence of the landscape as both natural formation and cultural construction. In this perspective, the landscape may be interpreted as a geographical space molded by natural forces. This can be subsequently transformed through human interventions, which contribute not only to its physical morphology, but also to its political, social, and cultural identities (Mosconi et al., 2020). Accordingly, the notions of landscape, territory, space, and environment, tend to converge, as the interweaving of human activity with the biotic and abiotic components of natural systems generates processes so deeply integrated that their boundaries become analytically indistinct.

The landscape is not a static entity. Rather, it is subject to continuous transformation, evolving across different temporal scales (Bastian et al., 2006). Natural dynamics such as erosion, vegetation growth, or climatic variation, interact with anthropogenic dynamics, including settlement, infrastructure, and land-use change, producing transformations that may unfold gradually or abruptly, and often through complex, overlapping trajectories (e.g., Paoletti et al., 2019). Within the field of geography, landscape analysis has traditionally occupied a central (theoretical) position. It has been regarded as a synthesis of natural and human dimensions of space, in which the visible form of territories embodies both its material structure and the network of relationships that animate it (Prokopová et al., 2018).

The course of history has marked a decisive turning point in the reflection on the concept of landscape, encouraging a critical reassessment of its traditional definitions (Renes, 2015). Geography itself has progressively broadened its scope, shifting from descriptive analysis to the examination of more urgent and applied questions—such as economic inequality, demographic growth, urbanization, and territorial transformations (Recanatesi et al., 2013). Within this (expanded) disciplinary framework, landscape studies acquire renewed significance, as they are intrinsically tied to cultural dimensions. The term *culture* is understood here as the ensemble of knowledge, values, and intellectual capacity across diverse domains that contribute to the individual's formation, both intellectually and morally (Akmaljonovich, 2021). This cultural component of the landscape allows individuals and communities not only to identify with their environment, but also to develop a critical awareness of social dynamics and ongoing transformations (Reynaud et al., 2020).

The notion of "culture" is polysemous and complex, often invoked when concrete or rational explanations fail to account for certain human behaviors, conflicts, or attitudes. It is frequently used to denote the physical and symbolic traits of nations' or ethnic groups' expressions such as "Hispanic culture" or "Western culture" exemplify this use. At the same time, culture encompasses artistic and intellectual production, including music, painting, theatre, and literature, as well as the forms of mass culture mediated through radio, television, and digital communication (Hofstede, 2001). In a broader sense, culture may also be understood as a linguistic and social competence, namely the capacity to preserve local dialects, acquire new languages, and engage with the traditions of other social or ethnic groups. Ultimately, culture represents the totality of human ways of living, including language, customs, religious practices, habits, and value systems.

From an anthropological perspective, it designates the ensemble of social, political, and spiritual manifestations that have historically shaped human existence. Traditions, customs, artistic expression, and political structures thus constitute essential components of both individual and collective identity—introduced, transmitted, and transformed by human agency. Just as these cultural values contribute to defining identity, they also play a formative role in shaping territories and landscapes. Enriched by the traces of history, landscapes embody the evolution of human presence and bears witness to both natural processes and anthropogenic transformations. They stand as an enduring record of human permanence in space—an indelible testimony of the reciprocal interaction between environment and culture.

Each geographical space can thus be read as the outcome of successive layers of civilizations and cultural expressions, sometimes complementary, sometimes heterogeneous. The continuous interaction among natural, cultural, and historical factors has shaped the distinctive physiognomy of landscapes in both advanced economies and emerging countries. This is particularly evident in a continent with long-lasting anthropogenic (settlement) processes such as Europe. Despite a marked heterogeneity, European landscapes remained profoundly influenced by historical processes. Their sociocultural connotations emerge as the cumulative outcome of the relationship between environmental components and the millennial culture that has inscribed their presence through enduring traces and documentary evidence.

Today, landscape transformations occur with increasing rapidity and continuity, primarily as a consequence of human activity (Rodrigo-Comino et al., 2021). Processes such as urbanization, industrial and infrastructural development, and the intensification of agricultural and pastoral systems have profoundly altered its original structure. The dynamism of social and economic development thus becomes a decisive factor in shaping territorial change and redefining the environmental framework in which landscapes evolve (Salvati, 2012). However, the concept of landscape understood as a collective heritage, an identity marker, and a historical testimony, is not always perceived or valued as such.

Even small local communities, in their efforts to improve social conditions, often become absorbed in the logic of technological progress and economic competitiveness, seeking advantageous positions in contexts that reward multifunctionality and innovation. In doing so, however, they risk losing the sense of care and responsibility toward their environmental and cultural contexts. The development of transport and communication networks facilitated daily mobility between urban centers and peripheral areas, promoting metropolitan expansion and fueling an ever-growing demand for new residential settlements (Salvati, 2013). Concurrently, the rise of intensive mass tourism, coupled with the delayed recovery of historical centers, has encouraged speculative exploitation of local territories. These phenomena exemplify broader processes of social transformation whose impact (abrupt and pervasive) has redefined settlement structures, modified lifestyles, and accelerated socioeconomic development.

The evolving relationship between cities and the surrounding landscape has produced fragmented territorial structures in which natural elements are increasingly overshadowed by expanding metropolises. Contemporary societies remain characterized by the legacy of uncontrolled industrial growth, which has contributed significantly to environmental degradation (Salvati, 2014). One of the most distinctive characteristics of modern landscapes, compared with more "traditional" territories, is its tendency toward architectural and economic standardization. Vast areas once devoted to agriculture or forestry have been converted into construction sites or intensive monocultures, leading to the gradual erosion of biodiversity and the loss of irreplaceable ecological wealth. The historical and cultural traditions of these regions embodied in their products, practices, and vernacular landscapes have long enriched territorial identity. Yet, under the dominance of economic systems increasingly oriented toward short-term profit, the traces of cultural memory and the signs of historical continuity risk being erased, leaving landscapes impoverished in both ecological and cultural terms.

Anthropogenic Landscapes

It is now essential to clarify what George Simmel (Kühne & Edler, 2022) means by the terms *nature* and *landscape*:

> *By nature we mean the infinite connection of things, the uninterrupted birth and destruction of forms, the fluctuating unity of events, which is expressed in the continuity of temporal and spatial existence.*

In these lines, Simmel encapsulates a notion of "nature" as a complex system in perpetual transformation, within which every element interacts and conditions the others. This dynamic conception of nature as an open, self-regulating process stands in marked contrast with that of landscape (Berger, 2023). According to Simmel, whereas nature represents an unbounded continuum, the landscape notion arises only when certain limits are introduced. The act of delimitation presupposes the presence of an observer: landscape is not merely a segment of nature but rather a cultural and perceptual construct, the outcome of a selective framing mediated by human consciousness. In this sense, landscape emerges at the intersection of the objective reality of nature and the subjective perception of the observer.

Yet, even as the observer isolates a portion of the natural continuum, that image remains inseparable from the broader environmental totality. For this reason, any attempt to establish precise boundaries between nature and landscape proves inherently problematic: the two concepts are interdependent, mutually constitutive, and inseparable within Simmel's theoretical framework (Waterton & Watson, 2014). This interrelation raises a fundamental question: to what extent can a landscape still be considered "natural?" Every image fixed in human memory is, by necessity, the image of a landscape already shaped directly or indirectly by human presence. From the cultivation of a wheat field to the construction of the tallest skyscraper, human intervention has continuously modified natural processes, leaving an indelible and, by definition, non-natural imprint upon the visible environment.

Within this context, the evaluation of human interventions becomes crucial. Both a wheat field and a residential building may either harmonize with the surrounding environment or disrupt its ecological and visual balance (e.g., Salvati, 2015). For instance, a wheat field managed through intensive agricultural practices often entails the removal of trees, hedgerows, and drainage canals, producing adverse effects on ecosystems and compromising the scenic coherence of the landscape. Conversely, small to medium-sized farms that adopt sustainable agronomic techniques such as crop diversification, soil conservation, and the maintenance of ecological corridors contribute positively to biodiversity and environmental resilience.

This qualitative distinction underscores the need to orient human activity toward more integrated and ecologically sensitive interactions with the environment, whether natural or anthropogenic. Policy failures are exemplified in the phenomenon of urban sprawl (Rubiera-Morollón & Garrido-Yserte, 2020), particularly evident in peri-urban areas and along the corridors connecting major urban centers. This unregulated expansion has generated hybrid landscapes, where rural and urban features coexist in fragmented and often incoherent configurations. Ensuring higher-quality forms of territorial transformation therefore requires spatial planning and policy regulation informed with a multidisciplinary analysis of ecological, social, and economic dynamics (Salvati, 2016). Nevertheless, over the past half century, patterns of territorial development have been largely dictated by economic speculation and political opportunism, rather than coherent planning or ecological foresight. As a consequence, land consumption has accelerated to alarming levels, leading to the progressive degradation of rural landscapes and the depletion of vital ecosystem services that constitute the foundation of natural capital.

Cultural Landscapes

History and anthropology have played a decisive role in delineating the characteristics and identity of landscapes. Through their methods and theoretical developments, these disciplines have profoundly influenced the geography of landscapes, shaping both interpretive frameworks and analytical tools. The traces of history reveal the cultural diversity of territories, making visible the processes that generate their heterogeneity and complexity over time. In this sense, the cultural landscape may be defined as the ensemble of elements produced by a specific cultural group within a given environmental context. Landscape transformations are thus inseparable from the cultural matrices that guide human behavior, making culture a central category of geographical reflection.

Carl O. Sauer, widely regarded as one of the founders of the modern concept of "cultural landscapes," provided an influential contribution to this perspective (Mathewson, 2025). Sauer emphasized that human culture is a fundamental agent in shaping the morphological characteristics of landscapes. In his view, culture acts upon the natural environment, transforming it into a meaningful spatial expression, namely a tangible manifestation of collective identity and creativity. Territories, in this framework, become the medium through which humanity expresses its values, practices, and symbolic systems; landscapes, in turn, reflect the identity and personality of the local community (Denevan & Mathewson, 2009). Humanity, therefore, does not merely occupy space; it continuously transforms in its own image and likeness. The elements composing the landscape constitute the medium through which culture materializes, and their interaction gives rise to the cultural landscape, intrinsically associated with physical and ecological contexts in which human activities unfold. As Vallega (2003) and Banini and Ilovan (2021) argue, cultural landscapes are fashioned from a natural landscape by a cultural group: culture is the agent, the natural elements are the medium, and the cultural landscape is the result.

A landscape rich in historical signs and material traces stands as a visible testimony to the cultural identity of humanity (Uzun & Somuncu, 2018). It serves as a palimpsest of successive civilizations and social groups that have inscribed their presence upon the territory, leaving behind (enduring) marks that make each place unique and multifaceted. Through the continuous turnover of generations, new cultural layers are superimposed upon older ones, producing a dynamic stratification of meanings and forms. Thus conceived, the cultural landscape functions as a living document of the past: it preserves the traces of human activity while also manifesting the shaping power of natural forces. Many of the most evocative landscape profiles bear witness to the interplay between natural and anthropogenic processes that have determined the origin, transformation, and persistence of places over time.

Preserving Cultural Heritage

The first concept to address when examining the relationship between heritage and landscape concerns the notion of "heritage" and its intrinsic value (Tilley, 2006). Traditionally understood as the wealth and potential of a given territory, the concept of heritage has progressively expanded in scope and meaning. The processes of globalization have profoundly transformed its perception, transcending the boundaries of local communities and fostering reciprocal exchanges among cultures and traditions on a global scale. As a result, contemporary societies are increasingly multiethnic and multicultural, where local folklore interacts sometimes harmoniously, sometimes contentiously with the cultural expression of other groups of non-native origin.

Over time, the notion of heritage has also extended into the natural domain—from historic gardens where architecture, art, and vegetation coexist harmoniously with natural parks and ecosystems, now considered integral components of collective patrimony (Pranskūnienė & Zabulionienė, 2023). To fulfill its social and cultural function, however, exploiting heritages requires not only the recognition of their value, but also the assumption of shared responsibility for its preservation. As a collective inheritance, heritage constitutes both "historical testimony" and "cultural identity," shaping the singularity of each community.

In this sense, heritage represents both inheritance and legacy: a continuum of historical values transmitted through time, reflecting who we were, who we are, and who we aspire to become (Graham & Howard, 2016). Safeguarding heritage therefore entails preserving history, interpreting its lessons, and addressing present challenges to prevent future social, economic, political, or cultural crises while simultaneously fostering endogenous and sustainable development. Every community builds its own cultural heritage, shaped by language, customs, and religious practices, as well as political and artistic expressions. These tangible elements, together with intangible aspects such as beliefs, rituals, and folklore, constitute the distinctive features of the landscape.

The contemporary understanding of cultural heritage thus embraces both material and immaterial dimensions (Harrison et al., 2020). Intangible heritage including oral traditions, rituals, artisanal practices, and popular culture, forms a living legacy of identity that complements tangible assets of historical, artistic, or archaeological significance. Landscapes in which social evolution intersects with natural elements are therefore recognized as cultural landscapes. United Nations Educational, Scientific and Cultural Organization (UNESCO) has played a pivotal role in redefining the boundaries of cultural heritage (Von Schorlemer, 2020). The 1989 Recommendation on the Safeguarding of Traditional Culture and Folklore (Ozer, 2023) established cultural diversity as a "common heritage of humanity," essential for creativity and innovation in the same way that biodiversity is vital for ecological resilience (Gkana, 2020). UNESCO has made the protection of intangible heritage a central objective, recognizing that cultural identity is transmitted through both material and immaterial expression (Rogers, 2023).

Cultural heritage, in this broader sense, manifests through practices that define human identity—languages and dialects, rituals and festivities, agro-pastoral traditions, and artisanal production (Andersen, 2025). These practices are deeply rooted in the landscape, molded by seasonal cycles, local resources, and geographical specificities. Traditional craftsmanship and agriculture, for instance, embody territorial identity and attest to the long-standing interaction between human societies and their environments. Although human action can lead to ecological degradation, it has also contributed to the creation of distinctive landscape systems that cannot be excluded from the definition of "cultural landscape." UNESCO recognizes as cultural heritage those places where natural and human elements coexist in mutual equilibrium, generating sites of exceptional beauty and universal value. Yet not every transformed environment qualifies as "heritage": only those landscapes in which cultural and natural components converge in distinctive, identifiable, and historically meaningful ways are deemed worthy of such recognition (Jakubowski & Schreiber, 2022).

A landscape becomes cultural when it exhibits unique and recognizable traits, including traditional settlements, monumental art, architectural achievements, or ecologically significant habitats. Together, natural and cultural heritage constitute a material and immaterial legacy that testifies to the history of human action within the territory, forming a public good that should be preserved and transmitted to future generations (Brașoveanu, 2024). Despite the extraordinary richness of historical and cultural landscapes, however, the resources allocated to their protection remain insufficient. This deficiency is particularly evident in rural and sparsely populated areas, which experience inadequate institutional support though often in more favorable conditions for habitat conservation than urbanized zones. Heritage, as both a testimony to the past and a foundation of collective identity, requires not only recognition, but also the implementation of effective, long-term policies for its protection, management, and enhancement.

Economic Valorization and Legal Protection of Cultural Heritage

In terms of public spending, it is important to underline that natural landscapes often function as a backdrop (or even a scenography) for human activity. The integration of natural and anthropogenic components generates what is defined as the cultural landscape (Oosterbeek, 2025), distinguished by geographical, historical, and social connotations associated with the place of origin, traditions, and customs. Both urban and rural landscapes, as products of human intervention, contribute decisively to shaping cultural heritage (Joldescu-Stan, 2023).

Human settlements, in particular, are characterized by the presence of a historic center, a symbolic and spatial node in which the community can recognize itself and reaffirm its collective identity. The enhancement of cultural heritage thus requires a renewed commitment to the restoration, recovery, and conservation of historic buildings and public spaces. This responsibility lies primarily with national and local authorities, as it concerns not only the protection of identity but also the sustainable regeneration of degraded or abandoned areas.

A key factor in defining the distinctiveness and attractiveness of urban centers lies in the presence of green areas and historical parks, often of notable artistic and botanical value. Such spaces enhance both the aesthetic and cultural dimension of cities, transforming them into focal points for residents and visitors alike. In many cases, these gardens and parks, whether managed as public goods by local administrations or maintained by private entities, collectively enrich the heritage, livability, and ecological quality of metropolitan environments. Conversely, rural areas represent one of the most fragile and undervalued components of both the landscape and national heritage (Lennon & Barrett, 2024).

They are among the most vulnerable territories, increasingly compromised by the excessive and often uncontrolled expansion of urbanization. The abandonment of agricultural land significantly contributes to the erosion of rural spaces (Knapik & Król, 2023). Yet, just as historical centers and monumental cities are recognized for their artistic and architectural significance, the countryside should equally be considered an integral part of cultural heritage. Its value resides in the persistence of traditional agricultural practices and crop systems that have historically shaped local identity. One of the most critical phenomena affecting the loss of arable land is the depletion of the soil's humic component, especially when agricultural areas are converted into urbanized zones. Such transformations often driven by intensive building activity on peri-urban green belts result in a dual loss: a reduction of cultivable land, and the erosion of cultural landscapes rooted in centuries of agrarian traditions.

These dynamics not only compromise the aesthetic quality and intrinsic value of places, but also deprive local communities of the basic resources necessary for agricultural productivity and economic sustainability (Bandarin & Van Oers, 2012). The progressive expansion of peri-urban areas into formerly rural zones erodes the perceptible distinction between city and countryside (Sanz Tolosana & González Fernández, 2020). Consequently, the agricultural sector suffers commercial and economic setbacks, while natural landscapes lose their heterogeneity, assuming the characteristics of a standardized and fragmented territory.

Urbanization is closely associated with land abandonment, which often triggers processes of spontaneous renaturalization that, while ecologically valuable, reduce the original productive and cultural value of rural land (Espín, 2022). This transformation also affects rural tourism, which depends on the authenticity and recognizability of traditional landscapes. Despite various European directives that promote the regeneration of abandoned urban areas and discourage further land consumption, advanced economies continue experiencing a steady increase in the extent of buildable land, designated not only for residential use, but also for commercial and industrial purposes.

The creation of new built environments entails the expansion of transport and communication infrastructures, which fragment the urban fabric and require extensive networks for water, energy, and electricity supply. These processes aggravate land consumption through the proliferation of impermeable surfaces, leading to a measurable decline in environmental quality and the disappearance of traditional agricultural practices. The loss of such practices corresponds to the fading of landscape's intrinsic diversity, functionality, and beauty.

A further challenge concerns legality and compliance with spatial planning and environmental regulations. Advanced societies continue bearing the legacy of a culture shaped by the unchecked industrial expansion of the last economic boom (Gillen, 2022), when widespread construction produced severe environmental degradation. The architectural consequences of that period remain visible today, representing a major cause of the deterioration of artistic and cultural heritage. These conditions underscore the importance of safeguarding quality of life in its environmental, landscape, and urban dimensions. Special attention should be directed to regions where illegality and corruption persists, as these practices undermine sustainable development and threaten both the cultural identity and the ecological integrity of entire regions.

Agricultural Landscapes

The agricultural landscape (i.e., the portion of land designated for cultivation and characterized by fields, pastures, and productive systems) represents one of the most enduring expressions of the interaction between human society and the environment (Mueller et al., 2021). Often referred to as the rural landscape, it has been extensively studied with regard to the historical evolution of agricultural practices that have shaped its present configuration. Transformations of the agrarian landscape are the result of long-term processes associated with social and economic dynamics, technological innovation, and cultural adaptation (Biasi et al., 2019).

Agriculture has profoundly shaped the landscape of highly anthropized regions, influencing not only morphology and land use but also social, cultural, settlement, and infrastructural structures. Following the definition of Sereni (1961), the agrarian landscape is envisioned as a form that humankind, in the course and for the purposes of its agricultural production activities, consciously and systematically, imprints upon the natural landscape (Ludwiczak et al., 2013). Sereni's approach emphasized the interdependence of history, human agency, and the natural environment, introducing methodologies that consider the landscape as a dynamic result of their continuous interaction. His study enriched by maps, artistic representations, and documentary sources traces the evolution of rural forms across successive epochs, each characterized by distinctive land management and settlement patterns. As Aimar (2024) noted, a complete understanding of contemporary landscapes would be impossible without such historical inquiry.

From the Neolithic through the Bronze and Iron Ages, early agricultural activities progressively intensified, marking the first significant human imprint on the natural environment (Dong et al., 2022). With Greek colonization and Etruscan synoecism, the political and social unification of preexisting villages into single centers allows systematic land cultivation. The introduction of fallow system transformed once-abandoned land into stable agricultural areas, establishing a decisive separation between natural and cultivated landscapes (e.g., Canfora et al., 2017). These transitions reflect an intricate balance between agricultural innovation, urban expansion, land abandonment, and renaturalization processes (Jakubínský et al., 2021).

Sereni devoted particular attention to the "Mediterranean gardens," describing them as complex agricultural systems characterized by the "predominance of rich and valuable trees and shrubs." According to his analysis, the "absolute prevalence of tree crops" defines the Southern-most zones of the Mediterranean, where climatic and environmental factors favored mixed cultivation. Within these landscapes, vines, olives, apricots, and almonds coexisted in enclosed plots, while the introduction of citrus species endowed the region with a renewed identity and aesthetic allure (Salvati et al., 2015). This arboreal mosaic transformed Southern agrarian landscapes into what Sereni described as a veritable "paradise."

Sereni's interpretation situates the Mediterranean garden within a broader framework of polycultural agro–silvo–pastoral systems, where agricultural land, woodland, and pastures coexist (Cruz & Sequeira, 2024). Such landscapes were historically shaped by diverse traditional practices: timber cutting, coppicing, use of branches for fodder, controlled burning, terracing, irrigation, and transhumance. These interwoven practices reflect a sustainable equilibrium between productivity and ecological integrity. For instance, the Roman "villa urbana," integrating agricultural and residential functions, introduced a new aesthetic and organizational sensibility toward cultivated space (Tally-Schumacher, 2020). However, the subsequent decline of ancient urban centers, compounded by economic instability, slave revolts, and social fragmentation, led to a collapse of agricultural activity (Kosmas et al., 2015). The remnants of ancient infrastructures and monumental works (once expressions of a "second nature" serving civic purposes) became the ruins of a disintegrating agrarian landscape. Colonization policies had redirected settlement toward the plains, systematically dismantling fortified hilltop communities in favor of more productive lowlands structured around the fallow-field system (Moss, 2020).

Yet, during the Middle Ages, waves of barbarian invasions and persistent insecurity prompted a return to mountainous and defensible sites (Thorndike, 2023). The resulting pattern of perched villages, associated with subsistence agriculture and nomadic pastoralism, became a defining element of the Medieval landscape (Wickham, 2005). Lowlands, often rendered unhealthy by marshes and malaria, reinforced this settlement inversion. Within these rural systems, fortified towns and castra proliferated, and even partially ruined cities consolidated their perimetral walls to ensure protection. The agrarian landscape of the early medieval city was characterized by enclosed fields, orchards, vineyards, and kitchen gardens (Goodson, 2021). Urban gardens, in particular, remained a constant feature of medieval centers, as documented in maps depicting dense networks of small green plots both within and around built-up areas (Yan et al., 2022). Their modest scale and strategic distribution reveal their function as productive gardens, intended as essential components of the economic, social, and cultural fabric of medieval communities.

Epilogue: Linking Traditional to Contemporary Landscapes?

The importance and value of modern agricultural landscapes are immediately evident in their primary function, that is, the production of goods and resources for human use (Hemberger et al., 2021). Yet this utilitarian perspective invites a broader reflection on how rarely we appreciate the full significance of agricultural landscapes, both as expressions of human history, and as active components of contemporary (territorial) identity (Vardopoulos et al., 2021). Within this perspective, the territory ceases to be a purely natural system and becomes an organized environment, structured by the actions of farmers who, through the delineation of fields and the adoption of closed-field systems, establish a new spatial order (Conzen, 2021).

These landscapes exemplify the coexistence of production and conservation, serving as repositories of biodiversity where human intervention interacts harmoniously with natural processes rather than disrupting them—an approach that resonates closely with the principles of sustainable development (e.g., Salvati, 2018). Beyond their productive dimension, agricultural landscapes hold intrinsic historical and cultural value, functioning as living archives of collective memory and as spaces of aesthetic and symbolic significance that embody the *genius loci*, or spirit of place (Arntzen & Brady, 2008). Their spatial organization, cultivation patterns, and agricultural practices reveal the evolution of social, economic, and environmental relations over time. The promotion of local products, short supply chains, and territorial brands further contribute to maintaining and transmitting their identity (Convery et al., 2012). The enhancement and preservation of agricultural landscapes thus constitute essential strategies for countering land consumption and mitigating the effects of urban sprawl. By supporting and revitalizing peri-urban agriculture, it becomes possible to reestablish connections between urban districts and rural systems (e.g. Salvati et al., 2008; Duan et al., 2022; Mace, 2022).

Transitional zones—where the historical, ecological, and cultural dimensions of the landscape intersect—can evolve into urban parks and community gardens, serving as spaces for environmental education, social inclusion, and the recovery of collective identity. Since the origins of agriculture, humanity has continuously sought to modify and domesticate the natural environment—plants, animals, and landscapes to meet its needs (Redlich et al., 2021). Over the centuries, these transformations have intensified to such an extent that ecosystems untouched by human activity have become increasingly scarce. As a result, countless plant and animal species have disappeared, revealing the ecological costs of uncontrolled exploitation.

Through scientific and ecological knowledge, humanity has gradually recognized that altering the delicate balance of nature ultimately jeopardizes the environmental conditions essential for survival. The protection of biodiversity, the rational use of resources, and the reduction of energy consumption therefore emerge as imperatives for contemporary societies (Wu, 2021). Within this context, planning and management of landscapes have become fundamental components of sustainable development and spatial governance. The European Landscape Convention (2000) played a decisive role in revitalizing interest in landscape studies across Europe, promoting a shared awareness of landscapes as dynamic systems shaped together by natural and cultural forces.

It also emphasized the importance of interdisciplinary approaches integrating historical, geographical, ecological, economic, and perceptual perspectives, aimed at comprehending landscape transformations, peculiarities, and degradation risks (Krauss, 2016). Ultimately, the intimate knowledge of landscape dynamics is indispensable for guiding human action (Worthman et al., 2024). Only through a coherent and responsible relationship with the environment—supported by appropriate tools, planning frameworks, and civic awareness—can societies pursue a model of development that is both culturally grounded and ecologically sustainable.

References

Aimar, F. (2024). Landscape: Concepts and features in the Italian and Chinese contexts. In *The resilience of cultural landscapes: Perspectives from UNESCO world heritage sites* (pp. 53–75). Springer Nature.

Akmaljonovich, K. J. (2021). The content and essence of the concept of 'intellectual culture'. *European Journal of Humanities and Educational Advancements, 2*(5), 15–18.

Andersen, C. (2025). A flexible blending: Cultural heritage and the concept of cultural diversity in UNESCO. *Global Perspectives, 6*(1), 142481.

Antrop, M., & Van Eetvelde, V. (2017). *Landscape perspectives*. Springer.

Arntzen, S., & Brady, E. (2008). *Humans in the land: The ethic and aesthetics of the cultural landscape*. Oslo Academic Press.

Bandarin, F., & Van Oers, R. (2012). *The historic urban landscape: Managing heritage in an urban century*. Wiley.

Banini, T., & Ilovan, O. R. (2021). Introduction: Dealing with territorial/place identity representations. In *Representing place and territorial identities in Europe: Discourses, images, and practices* (pp. 1–19). Springer.

Bastian, O., Krönert, R., & Lipský, Z. (2006). Landscape diagnosis on different space and time scales: A challenge for landscape planning. *Landscape Ecology, 21*(3), 359–374.

Berger, V. (2023). From spatial forms to perception: Reassessing Georg Simmel's theory of space. *The American Sociologist, 54*(1), 123–146.

Biasi, R., Brunori, E., Ferrara, C., & Salvati, L. (2019). Assessing impacts of climate change on phenology and quality traits of *Vitis vinifera* L.: The contribution of local knowledge. *Plants, 8*(5), 121.

Bozkurt, S. G., & Basaraner, M. (2024). Spatio-temporal investigation of urbanization and its impact on habitat fragmentation in natural ecosystems of Istanbul using Shannon's entropy and landscape metrics in GIS. *Environment, Development and Sustainability, 26*(10), 26879–26907.

Braşoveanu, F. (2024). Considerations on cultural and natural heritage and the universal obligation to protect and transmit it to future generations. *Ovidius University Annals, Series Civil Engineering, 26*(1), 1–10.

Canfora, L., Salvati, L., Benedetti, A., & Francaviglia, R. (2017). Is soil microbial diversity affected by soil and groundwater salinity? Evidences from a coastal system in Central Italy. *Environmental Monitoring and Assessment, 189*, 1–15.

Convery, I., Corsane, G., & Davis, P. (2012). *Making sense of place: Multidisciplinary perspectives*. Boydell & Brewer.

Conzen, M. P. (2021). *The making of the American landscape*. Routledge.

Cosgrove, D. E. (2008). *Social formation and symbolic landscape*. University of Wisconsin Press.

Cruz, J., & Sequeira, M. (2024). Food as territory: Gathering, production and distribution of food in the typo-morphological research of Saverio Muratori. In *Eating, building, dwelling* (pp. 50–60). Routledge.

Dempsey, K. E., & Wilbrand, S. M. (2017). The role of the region in the European landscape convention. *Regional Studies, 51*(6), 909–919.

Denevan, W. M., & Mathewson, K. (2009). *Carl Sauer on culture and landscape: Readings and commentaries*. LSU Press.

Dong, G., Lu, Y., Zhang, S., Huang, X., & Ma, M. (2022). Spatiotemporal variation in human settlements and their interaction with living environments in Neolithic and bronze age China. *Progress in Physical Geography: Earth and Environment, 46*(6), 949–967.

Duan, J., Cao, W., Zheng, Y., & Zhao, L. (2022). On the optimization landscape of dynamic output feedback: A case study for linear quadratic regulator. In *2022 IEEE 61st conference on decision and control (CDC)* (pp. 4040–4045). IEEE.

Egidi, G., Salvati, L., Falcone, A., Quaranta, G., Salvia, R., Vcelakova, R., & Giménez-Morera, A. (2021). Re-framing the latent nexus between land-use change, urbanization and demographic transitions in advanced economies. *Sustainability, 13*(2), 533.

Espín, M. (2022). Autonomous re-naturalization of cities in a context of degrowth. *Journal of Political Ecology, 29*(1), 75–93.

Forman, R. T. T. (2014). *Land mosaics: The ecology of landscapes and regions*. Cambridge University Press.

Forman, R. T., & Godron, M. (1986). *Landscape ecology*. Wiley.

Gillen, S. (2022). *Narrating the industrial city: Intellectuals and the Italian economic boom (1955–1965)*, Doctoral dissertation, University of Pennsylvania.

Gkana, A. (2020). Peoples' heritage or states' heritage? Sovereignty in the UNESCO mechanism for the safeguarding of intangible cultural heritage. *Brazilian Journal of International Law, 17*, 40.

Goodson, C. (2021). *Cultivating the city in early medieval Italy*. Cambridge University Press.

Graham, B., & Howard, P. (2016). Heritage and identity. In *The Routledge research companion to heritage and identity* (pp. 1–15). Routledge.

Harrison, R., DeSilvey, C., Holtorf, C., Macdonald, S., Bartolini, N., Breithoff, E., & Penrose, S. (2020). *Heritage futures: Comparative approaches to natural and cultural heritage practices*. UCL Press.

Hemberger, J., Crossley, M. S., & Gratton, C. (2021). Historical decrease in agricultural landscape diversity is associated with shifts in bumble bee species occurrence. *Ecology Letters, 24*(9), 1800–1813.

Hersperger, A. M., Grădinaru, S. R., Pierri Daunt, A. B., Imhof, C. S., & Fan, P. (2021). Landscape ecological concepts in planning: Review of recent developments. *Landscape Ecology, 36*(8), 2329–2345.

Hofstede, G. (2001). *Culture's consequences: Comparing values, behaviors, institutions and organizations across nations*. Sage.

Imbrenda, V., Quaranta, G., Salvia, R., Egidi, G., Salvati, L., Prokopová, M., Coluzzi, R., & Lanfredi, M. (2021). Land degradation and metropolitan expansion in a peri-urban environment. *Geomatics, Natural Hazards and Risk, 12*(1), 1797–1818.

Incerti, G., Feoli, E., Salvati, L., Brunetti, A., & Giovacchini, A. (2007). Analysis of bioclimatic time series and their neural network-based classification to characterise drought risk patterns in South Italy. *International Journal of Biometeorology, 51*(4), 253–263.

Jackson, J. B. (2022). The vernacular landscape. In *Landscape meanings and values* (pp. 65–81). Routledge.

Jakob, M. (2006). *L'émergence du paysage*. Infolio.

Jakubínský, J., Prokopová, M., Raška, P., Salvati, L., Bezak, N., Cudlín, O., Cudlín, P., Purkyt, J., Vezza, P., Camporeale, C., Daněk, J., Pástor, M., & Lepeška, T. (2021). Managing floodplains using nature-based solutions to support multiple ecosystem functions and services. *Wiley Interdisciplinary Reviews: Water, 8*(5), e1545.

Jakubowski, A., & Schreiber, H. (2022). Bridging intellectual property and cultural heritage law in the practice of international governmental organisations. In *Research handbook on intellectual property and cultural heritage* (pp. 340–357). Edward Elgar.

Johnson, M. S., Quinn, M. J., Jr., Wickwire, T., Buonagurio, J., & Williams, M. A. (2020). Do spatially explicit wildlife exposure models improve the estimation of risk for small mammals? Case study: Application of spatially explicit exposure model to small mammal exposures to lead in heterogeneous landscapes. *Integrated Environmental Assessment and Management, 17*(1), 259–272.

Joldescu-Stan, G. (2023). Economic perspectives and the role of technology in cultural heritage protection and sustainable development. *Online Journal Modelling the New Europe, 43*, 18–31.

Jones, M., Howard, P., Olwig, K. R., Primdahl, J. R., & Sarlöv Herlin, I. (2007). Multiple interfaces of the European landscape convention. *Norsk Geografisk Tidsskrift—Norwegian Journal of Geography, 61*(4), 207–216.

Kairis, O., Karavitis, C., Kounalaki, A., Salvati, L., & Kosmas, C. (2013). The effect of land management practices on soil erosion and land desertification in an olive grove. *Soil Use and Management, 29*(4), 597–606.

Knapik, W., & Król, K. (2023). Inclusion of vanishing cultural heritage in a sustainable rural development strategy: Prospects, opportunities, recommendations. *Sustainability, 15*(4), 3656.

Kosmas, C., Detsis, V., Karamesouti, M., Kounalaki, K., Vassiliou, P., & Salvati, L. (2015). Exploring long-term impact of grazing management on land degradation in the socio-ecological system of Asteroussia Mountains, Greece. *Land, 4*(3), 541–559.

Krauss, W. (2016). European landscapes: Heritage, participation and local communities. In *The Routledge research companion to heritage and identity* (pp. 425–438). Routledge.

Kreplin, H. N., Santos Ferreira, C. S., Destouni, G., Keesstra, S. D., Salvati, L., & Kalantari, Z. (2021). Arctic wetland system dynamics under climate warming. *Wiley Interdisciplinary Reviews Water, 8*(4), e1526.

Kühne, O., & Edler, D. (2022). Georg Simmel goes virtual: From 'philosophy of landscape' to the possibilities of virtual reality in landscape research. *Societies, 12*(5), 122.

Lanfredi, M., Egidi, G., Bianchini, L., & Salvati, L. (2022). One size does not fit all: A tale of polycentric development and land degradation in Italy. *Ecological Economics, 192*, 107256.

Lanfredi, M., Coluzzi, R., Imbrenda, V., Nosova, B., Giacalone, M., Turco, R., Prokopová, M., & Salvati, L. (2023). In-between environmental sustainability and economic viability: An analysis of the state, regulations, and future of the Italian forestry sector. *Land, 12*(5), 1001.

Lennon, J., & Barrett, B. (2024). Conserving an underappreciated heritage resource: The rural landscape. *Parks Stewardship Forum, 40*(3), 5–14.

Ludwiczak, Z., Benni, S., & Tassinari, P. (2013). The study of rural landscape at the farm scale: Changes in traditional signs and structures. *Journal of Agricultural Engineering, 44*(s2), 31–40.

Luginbühl, Y. (2007). Pour un paysage du paysage. *Économie Rurale, 297–298*(1), 23–37.

Luginbühl, Y., & Howard, P. (2016). *Landscape and sustainable development: The French perspective*. Routledge.

Mace, R. (2022). *Peri-urban agriculture and sustainable development in the 21st century*. Academic Press.
Mancino, G., Nolè, A., Salvati, L., & Ferrara, A. (2016). In-between forest expansion and cropland decline: A revised USLE model for soil erosion risk under land-use change in a Mediterranean region. *Ecological Indicators, 71*, 544–550.
Marchi, M., Ferrara, C., Biasi, R., Salvia, R., & Salvati, L. (2018). Agro-forest management and soil degradation in Mediterranean environments: Towards a strategy for sustainable land use in vineyard and olive cropland. *Sustainability, 10*(7), Article 2565.
Mariani, F., Zambon, I., & Salvati, L. (2018). Population matters: Identifying metropolitan sub-centers from diachronic density-distance curves, 1960–2010. *Sustainability, 10*(12), Article 4653.
Mathewson, K. (2025). Landscape and environmental change: Carl Sauer on land and life. In *Making geography matter* (pp. 87–103). Routledge.
McLean, K. (2021). Temporalities of the smellscape: Creative mapping as visual representation. In *Modern approaches to the visualization of landscapes* (pp. 217–245). Springer.
Mosconi, E. M., Colantoni, A., Gambella, F., Cudlinová, E., Salvati, L., & Rodrigo-Comino, J. (2020). Revisiting the environmental Kuznets curve: The spatial interaction between economy and territory. *Economies, 8*(3), 74.
Moss, G. (2020). *Hostile ground: Terrain, force, and control in the Roman empire* (doctoral dissertation). University of North Carolina, Chapel Hill.
Mueller, L., Eulenstein, F., Dronin, N. M., Mirschel, W., McKenzie, B. M., Antrop, M., & Poulton, P. (2021). Agricultural landscapes: History, status and challenges. In *Exploring and optimizing agricultural landscapes* (pp. 3–54). Springer Nature.
Olwig, K. R. (2019). The practice of landscape "conventions" and the just landscape: The case of the European landscape convention. In *Justice, power and the political landscape* (pp. 197–212). Routledge.
Oosterbeek, L. (2025). Beyond culture/nature divides: New approaches and tools for a cultural integrated landscape management. *Diversity, 17*(6), 436.
Ozer, Y. (2023). Recommendation on the safeguarding of traditional culture and folklore: Period of establishment and implementation. *Milli Folklor, 140*, 16–26.
Pallasmaa, J. (2012). *The eyes of the skin: Architecture and the senses*. Wiley.
Paoletti, E., Alivernini, A., Anav, A., Badea, O., Carrari, E., Chivulescu, S., Conte, A., Ciriani, M. L., Dalstein-Richier, L., De Marco, A., Fares, S., Fasano, G., Giovannelli, A., Lazzara, M., Leca, S., Materassi, A., Moretti, V., Pitar, D., Popa, I., Sabatini, F., Salvati, L., Sicard, P., Sorgi, T., & Hoshika, Y. (2019). Toward stomatal–flux based forest protection against ozone: The MOTTLES approach. *Science of the Total Environment, 691*, 516–527.
Parra-Sanchez, E., Pérez-Escobar, O. A., & Edwards, D. P. (2023). Neutral-based processes overrule niche-based processes in shaping tropical montane orchid communities across spatial scales. *Journal of Ecology, 111*(8), 1614–1628.
Phillips, A. (2014). The international context: The European landscape convention. In *Gardens & landscapes in historic building conservation* (pp. 309–320). Routledge.
Pranskūnienė, R., & Zabulionienė, E. (2023). Towards heritage transformation perspectives. *Sustainability, 15*(7), 6135.
Prokopová, M., Cudlín, O., Včeláková, R., Lengyel, S., Salvati, L., & Cudlín, P. (2018). Latent drivers of landscape transformation in Eastern Europe: Past, present, and future. *Sustainability, 10*(8), 2918.
Recanatesi, F., Ripa, M. N., Leone, A., Perini, L., & Salvati, L. (2013). Land use, climate and transport of nutrients: Evidence emerging from the Lake Vico case study. *Environmental Management, 52*(2), 503–513.
Redlich, S., Martin, E. A., & Steffan-Dewenter, I. (2021). Sustainable landscape, soil and crop management practices enhance biodiversity and yield in conventional cereal systems. *Journal of Applied Ecology, 58*(3), 507–517.

Renes, H. (2015). Historic landscapes without history? A reconsideration of the concept of traditional landscapes. *Rural Landscapes: Society, Environment, History, 2*(1), 2.

Reynaud, C., Miccoli, S., Benassi, F., Naccarato, A., & Salvati, L. (2020). Unravelling a demographic "mosaic": Spatial patterns and contextual factors of depopulation in Italian municipalities, 1981–2011. *Ecological Indicators, 115*, 106356.

Rodrigo-Comino, J., Salvia, R., Quaranta, G., Cudlín, P., Salvati, L., & Giménez-Morera, A. (2021). Climate aridity and the geographical shift of olive trees in a Mediterranean northern region. *Climate, 9*(4), 64.

Rogers, L. (2023). *The safeguarding and protection of intangible cultural heritage in international law and policy* (Doctoral dissertation), Deakin University.

Rubiera-Morollón, F., & Garrido-Yserte, R. (2020). Recent literature about urban sprawl: A renewed relevance of the phenomenon from the perspective of environmental sustainability. *Sustainability, 12*(16), 6551.

Salvati, L. (2012). The spatial nexus between population growth and land degradation in a dry Mediterranean region: A rapidly changing pattern? *International Journal of Sustainable Development and World Ecology, 19*(1), 81–88.

Salvati, L. (2013). Rural sprawl, Mykonian style: A scaling paradox. *International Journal of Sustainable Development and World Ecology, 20*(2), 109–115.

Salvati, L. (2014). Population distribution and urban growth in Southern Italy, 1871–2011: Emergent polycentrism or path-dependent monocentricity? *Urban Geography, 35*(3), 440–453.

Salvati, L. (2015). Lost in complexity, found in dispersion: 'Peripheral' development and deregulated urban growth in Rome. *Cities, 47*, 73–80.

Salvati, L. (2016). From sprawl to compactness and back: Population dynamics (1848–2011) and the economic structure of a Mediterranean city. *GeoJournal, 81*(2), 319–332.

Salvati, L. (2018). The "niche" city: A multifactor spatial approach to identify local-scale dimensions of urban complexity. *Ecological Indicators, 94*, 62–73.

Salvati, L., Zitti, M., & Ceccarelli, T. (2008). Integrating economic and environmental indicators in the assessment of desertification risk: A case study. *Applied Ecological and. Environmental Research, 6*(1), 129–138.

Salvati, L., Mavrakis, A., Colantoni, A., Mancino, G., & Ferrara, A. (2015). Complex adaptive systems, soil degradation and land sensitivity to desertification: A multivariate assessment of Italian agro-forest landscape. *Science of the Total Environment, 521*, 235–245.

Sanz Tolosana, E., & González Fernández, M. T. (2020). The contribution of mobilities to the social sustainability of rural areas in a context of crisis: Structural conditions, social diversity and inequalities. In *Crisis and post-crisis in rural territories: Social change, challenges and opportunities in southern and Mediterranean Europe* (pp. 31–49). Springer.

Selman, P., & Swanwick, C. (2010). On the meaning of natural beauty in landscape legislation. *Landscape Research, 35*(1), 3–26.

Strecker, A. (2024). Global landscape governance on the 20th anniversary of the European landscape convention. In *Cultivating continuity of the European landscape: New challenges, innovative perspectives* (pp. 17–26). Springer.

Talento, K. (2024). Landscape mutation. In *Sustainable buildings from marble waste heaps: A new methodology for landscape regeneration and reuse* (pp. 21–57). Springer Nature.

Talento, K., Amado, M., & Kullberg, J. C. (2019). Landscape—A review with a European perspective. *Land, 8*(6), 85.

Tally-Schumacher, K. J. (2020). *Cultivating empire in ancient Roman gardens: Unearthing the tangled relationship between plants and their gardeners* (Doctoral dissertation). Cornell University.

Thorndike, L. (2023). *The history of medieval Europe: The development of Europe and its civilization—From the decline of the Roman empire to the beginning of the sixteenth century*. DigiCat.

Tilley, C. (2006). Introduction: Identity, place, landscape and heritage. *Journal of Material Culture, 11*(1–2), 7–32.

Troll, C. (2007). The geographic landscape and its investigation. In *Foundation papers in landscape ecology* (pp. 71–191). Springer.

Uzun, F. V., & Somuncu, M. (2018). Cultural landscape: An evaluation from past to present. In *Handbook of research on methods and tools for assessing cultural landscape adaptation* (pp. 1–27). IGI Global.

Van Eetvelde, V., & Aagaard Christensen, A. (2023). Theories in landscape ecology: An overview of theoretical contributions merging spatial, ecological and social logics in the study of cultural landscapes. *Landscape Ecology, 38*(12), 4033–4064.

Van Putten, S., O'Meara, C., Wartmann, F., Yager, J., Villette, J., Mazzuca, C., & Majid, A. (2020). Conceptualisations of landscape differ across European languages. *PLoS One, 15*(10), e0239858.

Vardopoulos, I., Tsilika, E., Sarantakou, E., Zorpas, A. A., Salvati, L., & Tsartas, P. (2021). An integrated SWOT-PESTLE-AHP model assessing sustainability in adaptive reuse projects. *Applied Sciences, 11*(15), 7134.

Von Schorlemer, S. (2020). UNESCO and the challenge of preserving the digital cultural heritage. *Santander Art and Culture Law Review, 6*(2), 33–64.

Waterton, E., & Watson, S. (2014). *The cultural geographies of heritage*. Sage.

Wickham, C. (2005). *Framing the early middle ages: Europe and the Mediterranean, 400–800*. Oxford University Press.

Worthman, C. M., Cummings, C. A., & Lende, D. (2024). The landscapes of lives I: An action landscape approach to practices and the interface of individual and society. *Ethos, 52*(2), 149–165.

Wu, J. (2021). Landscape sustainability science (II): Core questions and key approaches. *Landscape Ecology, 36*(8), 2453–2485.

Yan, J., Van der Linden, S., Tian, Y., Van Valckenborgh, J., Strosse, V., & Somers, B. (2022). Characterizing garden greenspace in a medieval European city: Added values of spatial resolution and multi-temporal stereo imagery. *Remote Sensing, 14*(5), 1169.

Zonneveld, J. I. S. (1979). Physical geography in The Netherlands. *Erdkunde*, 1–10.

Chapter 4
Landscape Dynamics and Local Development

Abstract The present chapter examines, from historical, geographical, and economic perspectives, the relationship between landscape dynamics and local development, emphasizing their interdisciplinary complexity and their centrality to future sustainability. Following a theoretical discussion that distinguishes the concept of landscape from that of environment, and interpreting the former as a polysemic and dynamic construct, our contribution demonstrates how landscape acts simultaneously as a resource, a marker of cultural identity, and a driver of socioeconomic transformation. The analysis explores the principal domains of interaction. They include agriculture (which has evolved from traditional systems into agro-industrial models, with profound implications for biodiversity, land consumption, and technological dependence), urbanization (which, since the nineteenth century, has radically reconfigured landscapes by concentrating population and economic activity, while generating social, environmental, and health imbalances), and tourism (now a strategic sector of the global economy, capable of enhancing territorial value but also vulnerable to systemic crises such as the COVID-19 pandemic). Further sections address culture (understood both as the foundation of collective identity and as a factor of local development, with the landscape serving as a vehicle of memory, belonging, and social cohesion), climate change (which introduces unprecedented global challenges and requires integrated mitigation and adaptation strategies), bio diversity (considered both as a common good and an economic asset whose conservation involves complex cost–benefit and ethical dimensions), and social well-being, intrinsically linked to landscape perception and to conditions of equity, inclusion, and efficiency in the shared management of resources. Finally, the chapter underscores the urgency of multiscalar governance rooted in sustainability, equity, and efficiency, referencing key policy frameworks such as the European Landscape Convention. Landscapes should no longer be conceived merely as the passive background of human activity, but as strategic and dynamic interfaces integrating economy, society, culture, and nature—basically a fundamental dimension in steering the just transition toward sustainable and resilient models of local development.

Keywords Sustainability · Urbanization · Agriculture 4.0 · Cultural identity · Climate change

L. Salvati et al., *Landscape Economics*, Advances in Spatial Science,
https://doi.org/10.1007/978-3-032-25263-0_4

Introduction

Beyond disciplinary boundaries, the economic and cultural dynamics of societies profoundly influence how landscapes are both shaped and understood. The landscape exerts a far greater influence than is often recognized, shaping climate regulation, hydrological processes, water availability, and the provision of ecosystem services. A circular relationship thus emerges between these domains: the landscape not only reflects societies and their culture, but also actively conditions their development, contributing to the identity and historical trajectory of local communities. These interconnections underscore the necessity of acquiring systematic knowledge of the landscape and ensuring its preservation to guarantee a sustainable future in which nature, cultural heritage, and community identity coexist in harmony with the broader objectives of local systems, that is, the creation of economic value, while minimizing negative externalities (Araújo et al., 2021).

Achieving such a balance requires an integrated and interdisciplinary approach that fosters active human participation in landscape management and conservation. This engagement is essential not only for enhancing social and economic well-being but also, more critically, for safeguarding biodiversity (Meyer et al., 2020). Within this framework, scholars refer to the "metabolism of socioeconomic systems," denoting the set of processes that, through analytical study, enable the progressive reconstruction and adaptive maintenance of a given system. This "mutation" allows for a comprehensive understanding of the transformation of social systems in contrast to natural ones.

For this reason, the previous chapter developed a theoretical and applied framework introducing local systems as aggregates of interdependent elements characterized by non-linear developmental trajectories. Landscapes fit dynamically within such visions, representing the distinctive expression of socioecological complexity. It mixes environmental and economic dimensions of local systems, encompassing both structure (organization and configuration, as studied in ecology) and functions (land use, as investigated in applied economics). In light of these considerations, the present chapter offers a comprehensive examination of the sectoral relationships between landscape and economic activities, approached from a historical–geographical perspective, and focusing on the reciprocal influence between human action and economic dynamics that continuously reshape local territories (Salvati & Carlucci, 2013).

Landscape and Agriculture

The foundational relationship between humanity and the natural environment underwent a profound transformation with the advent of agriculture (Wells & Stock, 2020). This agricultural revolution represented far more than a change in subsistence strategies, marking instead a cognitive and practical breakthrough that reflects the recognition that human destiny could be shaped through the active management of natural resources rather than passive adaptation to them. Through the systematic cultivation of plant species and the domestication of animals, early societies established the first intentional mechanism for harnessing ecosystem processes for sustained nutritional and economic benefits (Vicente-Vicente et al., 2022).

Agriculture should therefore be defined not merely as a set of farming techniques, but as a complex socioecological system in its own right. Its evolution is intrinsically associated with a wide array of interdependent factors, including technological innovation, historical trajectories, and forms of social organization. As such, agriculture constitutes a dynamic and adaptive system, continuously transforming across spatial and temporal scales. This enhanced (human) capacity to manage and manipulate ecosystems through controlled cultivation, selective breeding, and the shift from opportunistic hunting to regulated animal husbandry accelerated the pace and scale of environmental transformation. Agricultural societies thereby initiated an irreversible, large-scale modification of the natural landscape, a process that has intensified across millennia (Salvati & Carlucci, 2014).

Following centuries of incremental development, a second, equally pivotal transformation emerged in the mid-to-late 1970s, particularly within advanced economies. This phase was marked by the mechanization and industrialization of production processes, leading to the consolidation of the agro-industrial system (Jackson, 2024). This system reflected the final transition of agriculture from land-based practices to a form of industrial production, where technological rationalization becomes the organizing principle of rural economies (Güldner et al., 2021).

Agro-industrial systems in advanced economies actually represent an anthropogenic ecosystem almost entirely optimized by technological intervention. Within this model, human agents exert systematic control over biological and physical processes, driven by imperatives of yield maximization, scale efficiency, and profitability. Such intense regulation has generated a wide range of heterogeneous and often contradictory effects on the landscape (Wang et al., 2022). The environmental and ecological consequences of this model vary substantially, revealing a stark contrast between the highly modified, often homogenized agro-landscapes of industrialized regions and diversified, ecologically resilient rural landscapes of less developed or geographically isolated areas. This asymmetry underscores that the global diffusion of the agro-industrial paradigm remains uneven.

In many regions of the Global South, as well as in geographically marginal territories of advanced economies, traditional agricultural systems persist, rooted in local ecological knowledge and resource limitations. These systems stand in contrast with the monocultural imperatives of globalized production, often preserving landscapes characterized by what may be termed as "neutral elements," that is, features that maintain high levels of biodiversity, functional ecological connectivity, and relative equilibrium compared with the intensely managed, resource-dependent agro-industrial environments. Thus, the global agricultural landscape presents a heterogeneous mosaic of local systems, reflecting the uneven rates, intensities, and spatial scales of human impact on the environment (Estrada-Carmona et al., 2022).

Among the multiple ecological consequences associated with the intensification of agricultural and agro-industrial models, the loss of biodiversity stands out as one of the most critical. This process directly undermines the essential principle of symbiosis between human activity and the intrinsic self-regulating mechanisms of nature. The higher the degree of biodiversity, the more structurally complex, functionally articulated, and ecologically resilient the landscape becomes, thereby enhancing its long-term sustainability. This correlation is hardly surprising, given that ecological diversity constitutes one of the primary and most quantifiable indicators of both the developmental level and the ecological health of a given environment, particularly within systems operating under natural biophysical constraints (Salvati & Carlucci, 2015).

Beyond species loss, a series of other degradation processes have profoundly influenced the functional involution of landscapes pushing them toward states of reduced productivity and diminished ecological integrity. Pivotal among these processes are soil degradation and the systematic deforestation of wooded areas (Buchadas, 2025). Soil degradation compromises land productivity through erosion, salinization, compaction, and nutrient depletion, processes often exacerbated by intensive monoculture practices that prioritize short-term yield over the maintenance of long-term soil fertility and ecosystem balance. Similarly, large-scale deforestation, historically and contemporaneously undertaken to convert forested areas into arable land (Seboka et al., 2024), produces a cascade of ecological disruptions: the loss of carbon sinks, destabilization of hydrological and climatic cycles, as well as habitat fragmentation. Together, these dynamics amplify biodiversity decline and reduce the landscape's ability to sustain essential ecosystem services.

At a broader, macro-structural level, these ecological shocks are associated with demographic factors. The continuous growth of human populations operates as a pervasive driver of anthropogenic pressure on terrestrial ecosystems (O'Sullivan, 2020). Population expansion increases the demand for food, water, and energy, intensifying the conversion of natural landscapes into productive (or urbanized) land, and further accelerating degradation cycles. This continuous population growth, associated with economic expansion and the finite capacity of the planet's ecosystems for resource renewal and waste absorption, represents one of the primary limits to sustainable global development. Rising demographic pressure translates directly into escalating demand for essential resources, particularly food, water, energy, and raw materials.

Meeting this growing demand inevitably entails the expansion of cultivated land, frequently at the expense of natural ecosystems, thereby inducing profound and often irreversible landscape transformations (Svoray, 2022). Such dynamics are typically accompanied by a parallel increase in water demand, exerting further strain on already scarce hydrological systems. Likewise, intensified energy consumption for industrial, agricultural, and domestic purposes and the persistent reliance on wood for construction and heating in developing economies amplify environmental stress. Although technological progress and the discovery of new energy sources can alleviate supply constraints, the ecological consequences of extraction and consumption remain substantial and unresolved.

In the European context, an additional determinant of landscape transformations has been the Common Agricultural Policy (CAP), launched in 1962. Initially conceived as a mechanism of "production aid," the CAP sought to ensure food security, income stability for farmers, and increased agricultural productivity across member states. However, early policy frameworks focusing on price guarantees and direct payments decoupled from ecological criteria incentivized intensive agricultural practices (Langlais, 2023), producing significant unintended consequences for land use and the environment. The main adverse outcomes of this incentive structure included (i) accelerated deforestation (namely, the conversion of marginal (and forest) land into arable fields to maximize subsidized output), (ii) excessive chemical inputs (i.e., the widespread use of synthetic fertilizers and pesticides, contaminating soil and water and disrupting local ecosystems), and (iii) landscape homogenization, resulting in the removal of hedgerows, wetlands, and natural buffers to enable mechanization, thereby reducing biodiversity and ecological connectivity.

Recognizing these environmental trade-offs, the CAP has undergone successive reforms especially since the early 2000s, aimed at reconciling agricultural productivity with ecological stewardship. The introduction of "cross-compliance" and, subsequently, "greening" measures related, even more strictly, financial aid to compliance with environmental, animal welfare, and land management regulations (Hristov et al., 2020). These reforms sought to promote sustainable management of soil, water, and biodiversity. Nevertheless, while these policy innovations represent meaningful progress, they mark only the initial phase of a broader transformation.

Achieving durable environmental outcomes through the CAP requires continuous monitoring, regulatory strengthening, and enhanced financial support for truly regenerative agricultural practices. Today, the trajectory of global economic development indicates a structural shift beyond the agro-industrial model toward advanced industrial systems. Within this paradigm, the capital, technology, and managerial logic of manufacturing and service sectors dominate agricultural processes. Consequently, farming loses much of its historical autonomy, becoming deeply dependent on industrial inputs, technological infrastructures, and external supply chains.

This transformation has indisputably enhanced productivity. The integration of industrial technologies has allowed farmers to achieve unprecedented efficiency, increasing yields while reducing unit costs and reinforcing competitiveness. Yet this success is two-edged. The same technological integration that drives productivity also engenders systemic vulnerabilities and financial fragility. The most immediate vulnerability arises from escalating capital intensity. Advanced machinery, precision technologies, and proprietary systems demand high upfront investment, ongoing maintenance, and technological renewal. This cost structure alters the traditional financial profile of farming and heightens exposure to market volatility in the global prices of industrial inputs (e.g., fuel, metals, patented seeds, agrochemicals). Moreover, technological lock-in restricts farmers' adaptive capacity, binding them to specific supply chains and software ecosystems that may not align with local ecological or market conditions.

Thus, while industrial integration has delivered significant productivity gains, it has also substituted biophysical risks (e.g., drought, floods, pests) with sociotechnical vulnerabilities — risks deriving from infrastructure dependency, energy supply, and systemic economic fragilities. For instance, precision agriculture systems dependent on uninterrupted electricity and digital connectivity can experience catastrophic (operational) failures in the event of energy shortages, natural disasters, or geopolitical disruptions. In such cases, the locus of control shifts from the individual farm to industrial and infrastructural intermediaries, revealing the embedded fragility of the modern agricultural-industrial nexus.

These structural dependencies are further compounded by environmental feedback loops that translate ecological degradation into economic instability. The overuse (or misapplication) of industrial technologies (whether chemical, mechanical, or digital) produces externalities that eventually re-enter the system as long-term costs. These are exemplified with three different processes, such as (a) the uneven decline in soil fertility (because continuous single-cropping and excessive chemical inputs deplete organic matter and micronutrients, undermining productive capacity), (b) reduced product quality (since environmental stress and genetic standardization diminish nutritional value and market differentiation), and (c) increased external costs (as far as pollution mitigation, habitat restoration, and regulatory compliance impose cumulative financial burdens).

In essence, the drive for short-term productivity under the industrial paradigm sacrifices ecological resilience for efficiency, generating a model that is both environmentally unsustainable and financially precarious. The critical challenge for twenty-first-century agricultural policy lies in decoupling technological progress from systemic vulnerability to foster innovation without compromising ecological (or socioeconomic) stability. Achieving this equilibrium requires a dual commitment: continuous technological advancement accompanied by robust ecological safeguards.

A promising technological paradigm in pursuit of this balance is "Agriculture 4.0" (Zhai et al., 2020). This model represents an evolutionary step beyond mechanization—an intelligent, data-driven system integrating traditional farming with digital technologies such as Internet of Things (IoT), artificial intelligence, big data analytics, and remote sensing (Araújo et al., 2021). These tools enable site-specific management of resources, allowing farmers to optimize the use of water, fertilizers, and pesticides while minimizing waste and environmental impact. By harnessing real-time data on soil chemistry, microclimate, pest dynamics, and crop health, precision agriculture facilitates proactive, evidence-based decision-making. The result is not only enhanced productivity, but also a potential reduction in the ecological footprint per unit of production.

However, the success of Agriculture 4.0 critically depends on the human dimension. As noted by Javaid et al. (2022), digital literacy, technical competence, and interpretive capacity among farmers determine whether these technologies become tools of empowerment or sources of dependency. Without adequate training, knowledge transfer, and institutional support, even the most advanced systems risk underutilization, misuse, or counterproductive complexity. Therefore, the transition toward a sustainable agricultural future should rest on a dual foundation: technological innovation and investment in human capital. Only through this synthesis can the agricultural sector effectively decouple productivity from ecological degradation, achieving a resilient, knowledge-based model of sustainable development that aligns economic viability with the imperatives of environmental stewardship.

Landscape and Urbanization

"Urbanization" refers to the progressive transformation of territories into urban areas, characterized by high population density and the concentration of economic, social, and cultural functions. This process entails the continuous expansion of infrastructure and services, producing a profound reshaping of natural landscapes. Agricultural and rural spaces are gradually converted into residential settlements, industrial plants, public facilities, and transportation networks. Urbanization thus represents not merely a physical modification of space, but a structural reorganization of societies defining how communities interact, exchange, and evolve. The effects of urbanization extend far beyond the built environment.

Cities, as complex socioeconomic systems, concentrate opportunities for employment, education, and cultural innovation, while simultaneously generating new challenges in sustainability, resource management, and social equity. Historically, urban centers have fostered cultural pluralism and technological progress, functioning as engines of modernization and economic development. Yet, the same process has also been accompanied by environmental degradation, public health risks, and social marginalization shaping the lived experience of contemporary urban life (Kudryavtseva, 2020).

Historically, urbanization accelerated during the nineteenth century, coinciding with the Industrial Revolution. The symbiotic relationship between cities and industry emerging during this time period became a durable structural bond (Knowles et al., 2020). Industrial development required a spatially dense labor force, efficient transportation systems, and proximate markets, while cities benefited from the employment opportunities and technological innovations generated by industrial activity. This mutual dependence created a self-reinforcing feedback loop, intensifying the spatial and demographic expansion of cities, and consolidating their status as the main loci of modern economic and social transformation.

In this perspective, urbanization may be understood as both a driver and a mirror of modernization, continuously shaping landscapes and redefining territorial hierarchies. The close interconnection between industrial development and urban expansion facilitated the rapid improvement of transport infrastructure, enhancing mobility of goods and people. This transformation encouraged large-scale rural-to-urban migration, as populations once engaged in agriculture relocated to cities in search of industrial employment. Urban growth generated substantial benefits including increased productivity, a diversified labor market, and expanded access to services and amenities (Sorensen, 2020). Cities thus emerged as focal points of opportunity, promising upward mobility and improved living standards.

However, these benefits were accompanied by serious (structural) challenges. The rapid pace of metropolitan expansion frequently outstripped planning capacity, resulting in overcrowded districts, inadequate infrastructure, and declining living conditions. Public health crises emerged early in industrial cities, where high population densities coincided with poor sanitation and limited access to clean water. These vulnerabilities persist today: pandemics demonstrated how densely populated urban areas can become epicenters of contagion. Environmental degradation represents another enduring consequence of urbanization. Air and water pollution, traffic congestion, and the proliferation of impermeable surfaces have reduced the ecological resilience of local systems. Social fragmentation and insecurity also intensify within large metropolitan contexts. Rising crime rates and persistent inequalities contribute to the marginalization of vulnerable groups, including those experiencing poverty, structural unemployment, class or ethnic segregation.

Urbanization should therefore be interpreted as a double-edged process that stimulates economic growth and cultural diversity while simultaneously generating new forms of environmental and social vulnerability. A central challenge for contemporary societies lies in reconciling these contradictions through policies grounded in sustainability, inclusiveness, and spatial justice. Before 1840, the circulation of goods and people relied predominantly on carts and horses. As a result, cities remained spatially compact, with limited mobility reinforcing a form of social proximity: the wealthiest classes occupied urban cores, while socioeconomic status declined toward the periphery. The advent of new transport technologies in the second half of the nineteenth century—most notably horse-drawn carriages and electric trams revolutionized mobility and reshaped settlement morphology (Knoll & Reith, 2020).

Between 1850 and 1900, these innovations enabled the emergence of streetcar suburbs, where middle- and upper-class populations relocated to less congested, greener districts without sacrificing access to the city center. The early twentieth century introduced further transformations. Between 1900 and 1920, advances in communications and transport including the diffusion of telephones, motorized vehicles, and delivery vans facilitated the decentralization of production and services (Kaplan & Holloway, 2024). Economic activities increasingly moved toward the periphery, laying the foundations for polycentric and functionally specialized regional systems that characterize modern metropolitan regions.

A decisive turning point coincided with the mass production of cars in 1908 (Mom, 2020). Cars became both a symbol and an instrument of individual freedom, offering unprecedented flexibility of movement. Unlike collective transport, private automobiles allowed citizens to travel independently of routes and schedules, reinforcing a culture of personal mobility. The steady improvement of road infrastructure further amplified this autonomy, enhancing accessibility while drastically reducing travel times (Pillai et al., 2020).

Yet this mobility revolution introduced profound contradictions. The preference for private vehicles led to chronic infrastructural congestion: expanding road networks consumed vast tracts of urban land, often at the expense of green and public spaces. Parking facilities proliferated, eroding urban livability and transforming the visual and ecological character of cities. By the mid-twentieth century, car-oriented development had redefined the spatial organization of urban areas, influencing patterns of settlement, social interaction, and environmental quality (Smil 2021). Today, the cumulative effects of these transformations are visible in global demographic trends: over half of the world's population now lives in urban areas, with industrialized nations exhibiting mature urban systems and developing regions experiencing rapid metropolitan growth (Table 4.1). Cities have thus become the central arena of contemporary civilization concentrating political, economic, and cultural power, while simultaneously serving as laboratories of sustainability, inequality, and innovation.

Cities can be effectively defined as inherently complex socioenvironmental systems, where divergent forces, structural challenges, and transformative opportunities, continuously converge and interact. It is precisely this confluence of opposing dynamics that, beginning in the 1970s, stimulated a profound academic debate, seeking a more comprehensive conceptualization of the regional question. Among key contributors, Manuel Castells emphasized the necessity of understanding cities within the historical specificity of capitalist production systems, arguing that urban forms and functions are basically inseparable from the logic of profit and the process of economic production and valorization (Sevilla-Buitrago, 2022). For Castells, urban space is not merely a neutral container of social relations, but a structural expression of uneven economic and political forces.

Conversely, Peter Saunders (as discussed in Brenner, 2016) advocated for expanding the analytical framework beyond the strict limits of urban cores. He argued that the dynamics of urbanization cannot be understood in isolation from their surrounding territories, and that rural–urban interdependencies and regional-scale processes play a decisive role in determining socioeconomic outcomes. This

Table 4.1 Urbanization of the global population over the last three centuries (selected years). Source: Authors' elaboration of official statistics and web resources

Year	Population (millions)	Urban population (millions)	Urbanization rate (%)	Population change (%)
1700	102	12.6	12.3	–
1750	120	14.7	12.2	0.3
1800	154	18.6	12.1	0.5
1850	203	38.3	18.9	1.5
1880	243	71.4	29.3	2.1
1900	285	108.3	37.9	2.1
1910	312	127.1	40.8	1.6
1930	333	159.7	47.9	1.1
1950	367	186.0	50.7	0.8
1970	427	271.8	63.7	1.9
1980	456	304.1	66.7	1.1

broader spatial perspective redefined cities as part of a continuous urban field, in which flows of labor, resources, and capital transcend traditional administrative boundaries. Each local context possesses a distinctive constellation of historical, geographical, and socioeconomic characteristics that should be rigorously considered when formulating strategies for development and governance. Ignoring even a single one of these defining variables, or attempting to impose uniform policy models across diverse contexts, inevitably produces ineffective outcomes, compromising long-term sustainability and social well-being of both urban and rural communities.

With globalization, cities are increasingly interconnected, embedded within vast transnational networks of production, information, and migration (Ezeudu & Chukwudubem, 2024). This interdependence heightens the need for policy frameworks that are flexible, adaptive, and capable of responding to the unique challenges and potentials of each settlement. Ultimately, a critical risk persists today: the tendency to overlook the granular specificities that define each urban environment. Such oversight undermines the capacity to secure truly fair, inclusive, and sustainable trajectories of growth both for cities themselves and for the broader territories that sustain them.

Landscape and Tourism

Tourism is formally defined by the World Tourism Organization (UNWTO) as

> *the set of activities of persons traveling to and staying in places outside their usual environment for not more than one consecutive year for leisure, business and other purposes* (UNWTO, 2005–2007).

This comprehensive definition underscores that tourist movement arises from a wide spectrum of motivations, encompassing leisure, cultural and religious pursuits, gastronomic exploration, appreciation of natural environments, and objectives related to individual health and well-being. Within academic discourses, it is essential to distinguish between tourists and travelers. Tourists are typically defined as individuals whose principal motivation for visiting a destination is centered on leisure, relaxation, or the consumption of established, popular attractions (Canton, 2021). In contrast, travelers seek a deeper, more authentic engagement with the destination. Travelers experience local culture firsthand, interact meaningfully with the environment, and avoid the often-superficial layers of mass tourism infrastructure.

Although tourism has ancient historical roots, its transformation into a major global economic sector is a relatively recent phenomenon (Kala, 2021). The modern era of tourism can be traced back to Thomas Cook's entrepreneurial initiatives in 1845, which marked the birth of organized travel. The subsequent advent (and diffusion) of air transport revolutionized the industry, enabling large-scale mobility and giving rise to mass tourism—a form characterized by the simultaneous concentration of large numbers of visitors in specific destinations during the same period. This model often led to service standardization and imposes substantial ecological and social pressures on host communities.

Over the past two decades, tourism has shifted from a relatively elitist activity to a widespread global practice, engaging a significant proportion of the world's population. This democratization of travel has radically transformed tourism demand, which now centers on the pursuit of distinct, memorable, and emotionally resonant experiences. As a result, tourism enterprises should increasingly respond to diversified and sophisticated expectations that extend far beyond the provision of standard goods and services. Contemporary tourists seek immersion, authenticity, and values that compel the industry to prioritize experience creation over standardized production. The tourism sector employs several fundamental classifications to categorize movement patterns, distinguishing three main forms based on the geographical relationship between origin and destination (Fang, 2020):

- Incoming (inbound) tourism: activities undertaken by non-resident visitors arriving from other countries or regions to stay in a destination outside their usual environment
- Outgoing (outbound) tourism: travel by residents of a country to foreign destinations
- Domestic tourism: travel within a single country, where residents visit destinations outside their usual environment but without crossing national borders

These three forms of tourism generate distinct economic impacts on national and regional economies. Incoming tourism is often the most economically advantageous, as it constitutes a direct inflow of foreign currency into the host nation's economy. Visitors' expenditures on accommodation, food, transport, retail, and leisure directly strengthen the balance of payments, stimulate employment, and boost profits for businesses across tourism ecosystems. Outgoing tourism, conversely, represents an outflow of domestic expenditure toward foreign economies, thus an economic leakage for origin countries. Nevertheless, it indirectly contributes to domestic development by fostering exposure to new service standards, stimulating innovation, and supporting the expansion of airlines, travel agencies, and logistic infrastructures (Llorca-Rodríguez et al., 2020). Domestic tourism retains expenditure entirely within the national economy, providing a strong multiplier effect across local businesses such as hotels, restaurants, attractions, and event organization. Beyond its financial significance, domestic tourism is crucial for the valorization and preservation of cultural and natural heritage, as it generates local demand for conservation and sustainable management (Athari et al., 2023).

It is also important to recall that, according to UNWTO definitions, tourism explicitly excludes travel undertaken for the purpose of remunerated employment within the visited destination. The purpose of the visit should therefore differ from the pursuit of income-generating activities (Veal, 2025). This distinction is critical for separating genuine tourism from temporary labor migration or work-related mobility. The economic weight and societal impact of the tourism sector vary significantly from one region to another. This variation arises from a diverse array of pull factors and environmental attributes that collectively determine a destination's attractiveness and carrying capacity. These determinants include:

- Geographical and environmental assets, such as proximity to maritime or mountain areas, which underpin the potential for natural, adventure, or eco-tourism
- Cultural and heritage resources, including density, quality, and conservation status of monuments, museums, and historical sites
- Event-driven demand, namely the recurrence of major historical or cultural festivals, the organization of international conferences, and the presence of large-scale entertainment facilities such as amusement or theme parks

Central to the organization and maximization of these resources are tourist districts—distinct geographical areas characterized by a high concentration of specialized services, interconnected enterprises, and readily accessible attractions. The strategic purpose of developing such districts lies in achieving synergistic effects: by clustering resources, competencies, and infrastructures, these systems enhance competitiveness and stimulate sustainable (local and regional) economic growth more effectively than isolated enterprises could achieve independently. Within these districts, dense spatial and social interrelations that connect the economic dimension with local community life have demonstrated to strengthen local competitiveness (Trusova et al., 2020).

These relationships enable tourism enterprises to integrate deeply into their territorial context, fostering a nuanced understanding of local culture, visitor expectations, and emerging market niches. This embeddedness constitutes a key form of competitive intelligence and a fundamental advantage of spatial proximity. Such environments naturally encourage the formation of strategic partnerships among tourism enterprises, public institutions, regional planning agencies, and trade associations. These cooperative networks promote shared infrastructure, joint marketing, and pooled expertise, thereby reinforcing the entire economic fabric of the district (Burkaltseva et al., 2020). Moreover, by leveraging these spatial relationships, firms can assert the authenticity intrinsically associated with their operational landscape.

Drawing upon distinctive local heritage, environmental resources, and cultural identity, allows them to achieve powerful market differentiation; a stronger international profile authenticity becomes, in this sense, a core competitive asset. The aggregation of tourism activity within coordinated districts has also generated profound transformations in urban and territorial organization. Tourism stimulates demand for improved infrastructure, skilled labor, and high-quality services, thus fostering investment in innovation and planning. As a result, the sector plays an increasingly active role in shaping economic and territorial development strategies, becoming a pivotal component of sustainable regional development. Addressing this complexity requires a multidisciplinary approach to the management of natural, cultural, and landscape resources that balances competitiveness with environmental and social sustainability.

The effective fulfillment of this role depends on the continuous and innovative training of personnel, ensuring that skills remain aligned with rapid technological and market transformations. This evolution in human capital is essential to successfully manage the growing number of tourism and cultural projects that underpin local and regional economies. In this context, the COVID-19 pandemic represented an unprecedented global shock. The crisis affected not only public health and international mobility but also the social fabric and economic foundations of the tourism industry. Although historically characterized by high resilience and adaptive capacity, the sector's near-total shutdown revealed its profound vulnerability to exogenous shocks, emphasizing the urgency of risk management and diversification strategies for future stability (Jeon & Yang, 2021).

The impact was uneven across geographical and functional typologies: the hardest-hit segments included large metropolitan destinations and mountain municipalities reliant on seasonal tourism and international mobility (Salvati & Carlucci, 2016). By contrast, smaller-scale, nature-based or proximity tourism in lake and coastal regions, experienced more moderate declines, reflecting a temporary shift in tourist preferences (Fedyk et al., 2022). Employment levels in the tourism sector remain below prepandemic benchmarks across many regions. The first significant rebound occurred in early 2022, coinciding with the relaxation of international mobility restrictions. This resurgence was largely driven by leisure and wellness-oriented travel, reflecting a renewed prioritization of personal well-being following prolonged social confinement (Vărzaru et al., 2021).

In conclusion, the widespread border closures, cancellation of major events, and global travel restrictions inflicted severe economic damage, compounded by widespread layoffs and the bankruptcy of several small and medium-sized enterprises. Consequently, the recovery of the tourism sector remained slow and uneven, and its full return to precrisis levels may require several years. This structural fragility underscores the necessity of strategic policies focused on diversification, resilience, and sustainability, ensuring that tourism can continue to serve as both an engine of economic development and a steward of environmental and cultural resources (Duro et al., 2021).

Landscape and Culture

Cultural identity is understood in a broad and systemic sense as the ensemble of shared values, traditions, languages, beliefs, and symbolic expressions that uniquely define a local community. By elevating culture and heritage from the status of historical remnants to that of active strategic assets for sustainable development, it can be possible to transform landscape protection into a binding principle of public policies that aim to reconcile economic growth with environmental stewardship and the conservation of the historical, aesthetic, and cultural character of the national territory. This framework finds further support and amplification within the international legal order, particularly through the instruments adopted under the auspices of UNESCO. The organization's conventions most notably the 1972 Convention Concerning the Protection of the World Cultural and Natural Heritage, the 2003

Convention for the Safeguarding of the Intangible Cultural Heritage, and the 2005 Convention on the Protection and Promotion of the Diversity of Cultural Expressions—constitute the main (multilateral) mechanisms for the global protection and valorization of cultural diversity. Their adoption reflects a shared international consensus on the need to integrate the protection of heritage and identity into the broader agenda of sustainable development.

Specifically, the 2003 Convention for the Safeguard of Intangible Cultural Heritage (Heritage & Rii, 2020) and the 2005 Convention on the Protection and Promotion of the Diversity of Cultural Expressions (Lazaro Ortiz & Jimenez de Madariaga, 2022) represent the principal international reference points for the protection and promotion of cultural diversity. While these conventions provide robust operational mechanisms for global cultural governance, they also reveal definitional limitations when attempting to fully encompass the notion of "*anthropological culture*." In the academic sense, anthropological culture refers to the comparative and interpretive study of the multiplicity of human cultures across the globe, seeking to identify both their unique traits and shared patterns through a deep understanding of social practices, beliefs, and symbolic systems.

The challenge for global policy instruments lies in developing definitions of culture that are universally applicable and operationally precise, without erasing the contextual richness and relational complexity that anthropological practice seeks to preserve. This tension between the policy need for clarity and the anthropological need for nuance remains a persistent issue in the realm of global cultural governance. For instance, while the 2003 Convention for the Safeguarding of the Intangible Cultural Heritage provides an essential framework for preserving traditions, expressions, and knowledge systems, it tends to narrow the definition of cultural identity, often emphasizing artistic and performative dimensions at the expense of deeper social, epistemic, and relational components.

This limited focus risks reducing culture to a set of marketable artifacts or spectacles, rather than acknowledging it as a living, dynamic system of meanings, practices, and relationships. By contrast, the 2005 Convention on the Protection and Promotion of the Diversity of Cultural Expressions adopts a more evolutionary and process-oriented perspective. Often referred to as a "document of ten keys," it articulates ten guiding principles that collectively underscore the importance of protecting and promoting cultural diversity in a globalized world (Wiktor-Mach, 2020). This Convention explicitly recognizes culture as an adaptive process that evolves in response to contemporary challenges and social transformations (Patel, 2023). Despite their differing emphases, both conventions converge on a critical issue: the protection of Indigenous Intellectual Property (IIP).

This concept foregrounds the importance of intangible cultural heritage including traditional knowledge, oral traditions, folklore, designs, and community-based practices as a fundamental expression of collective identity. In this light, national and international policy frameworks should move beyond the mere preservation of heritage to address questions of ownership, benefit-sharing, and ethical use of traditional knowledge, particularly in the context of commercialization and global cultural exchange (Cheng & Yuan, 2021). This evolving understanding of culture profoundly influences contemporary planning and territorial governance.

The concept of "*cultural armor*" captures this transformation, describing culture as a strategic tool for territorial planning and protection. It signifies a shift from viewing the territory as a static, two-dimensional (geographical) entity to understanding it as a complex cultural system—an integrated matrix of natural assets, historical stratifications, and living traditions. Such a holistic view has become essential to modern planning, which increasingly seeks to integrate cultural vitality directly into regional development and sustainable resource management strategies. This paradigm shift reveals two complementary approaches to heritage management, namely the "protectionist approach," focused on safeguarding tangible and intangible assets through conservation and regulatory mechanisms, and the "prospective approach," which conceives of heritage as a dynamic resource for innovation, identity construction, and long-term strategic investment.

The "prospective" model emphasizes that heritage should inform and guide future development trajectories, ensuring that growth remains consistent with cultural and environmental contexts. Empirical evidence on the economic impact of cultural activities confirms the strategic importance of culture, underscoring the need for a more precise economic valorization of landscapes as generators of both material and immaterial wealth for local communities. In this expanded perspective, the landscape transcends its aesthetic function to become a medium of dialogue and cooperation among diverse cultural identities. The perception of landscapes is, however, inherently complex, encompassing two interrelated dimensions, the "identity" dimension and a "temporal sedimentation" dimension. Identity or "subjective perception" reflects the affective and symbolic connection that individuals (or communities) establish with their environment, through which they recognize themselves and their shared belonging. Otherness or "temporal sedimentation" reflects the objective, historical dimension of the landscape, representing the accumulation of natural and anthropogenic transformations over time—thus reflecting a layered record of collective memory that often exceeds human control or comprehension.

Understanding this duality is essential for developing policies capable of balancing subjective belonging with the objective temporal and ecological dynamics of place. Within sociological discourse, the concept of "*otherness*" (Romero-Rodríguez et al., 2021) is used to denote the specific (social, cultural, or ethnic) diversity that defines the intrinsic relationship between individuals and groups within a given territorial context. From this analytical standpoint, landscapes function as both a product and a producer of identity, shaping cultural practices and worldviews, while reflecting the evolving interaction between human and natural systems (Dangler, 2021).

Local landscapes, in particular, constitute the immediate environment in which a population is rooted and from which it draws its cultural references, values, and identity. It serves as both a repository of historical experience and a living framework of meaning. Within this context, commemorative practices, festivals, traditions, religious or civic anniversaries—play a fundamental role in reinforcing collective identity. These cyclical rituals are not mere leisure events; they function as mechanisms of cultural transmission, strengthening social cohesion and ensuring intergenerational continuity of shared values.

Beyond such practices, national symbols (anthems, flags, monuments, and memorial sites) anchor collective memory within physical space, embodying the intersection of identity, history, and territory. Yet in today's globalized context, marked by unprecedented mobility and connectivity, traditional boundaries of identity are increasingly "porous." Individuals now interact with (and absorb) influences from diverse cultural and territorial systems, leading to the emergence of multilayered and hybrid identities that enrich social diversity, but also risk diluting the rootedness of local cultures. Such dynamics underscore the urgent need to preserve and promote the diversity and authenticity of local traditions as a foundation for sustainable identity formation.

Cultural identity is therefore not merely a symbolic or sociological construct but a strategic driver of economic and social development. By fostering distinctiveness, it enhances territorial attractiveness, stimulates tourism, supports creative industries, and strengthens community cohesion. Investing in the preservation and renewal of cultural identity thus represents a core prerequisite for inclusive, resilient, and a sustainable growth that aligns economic vitality with the ethical imperative to safeguard the diversity and integrity of human cultures.

Landscape and Climate Change

Global warming, according to the Intergovernmental Panel on Climate Change (IPCC)—the leading global authority for assessing scientific data and risks associated with the climate system—is defined as

> *any change in climate over time, whether due to natural variability or as a result of human activity* (Solomon, 2023).

This definition is of fundamental importance, as it explicitly acknowledges that anthropogenic activities (industrialization, deforestation, and fossil fuel combustion) can significantly alter the chemical composition of the atmosphere, thereby triggering climatic variations that extend beyond the scope of natural processes. The global policy response to this emerging environmental crisis began formally in 1980 with the First World Climate Conference, which sought to coordinate international strategies aimed at mitigating the negative effects of industrial activity on the Earth's climate (Meehl et al., 2007). A few years later, the analytical work carried out by the International Commission on Climate Change which subsequently evolved into IPCC confirmed a measurable increase in average global temperatures, providing scientific validation for what is now recognized as anthropogenic climate change. This scientific consensus laid the groundwork for the United Nations Framework Convention on Climate Change (UNFCCC), adopted in 1992. The UNFCCC established the ultimate objective of

> *stabilizing greenhouse gas concentrations in the atmosphere at a level that would prevent dangerous anthropogenic interference with the climate system.*

It provided the foundational legal and institutional framework for all subsequent global climate agreements. The Conference of the Parties (COP), established under the UNFCCC, serves as the supreme decision-making body responsible for reviewing implementation progress, negotiating commitments, and determining the global trajectory of climate governance. The Conference of the Parties (COP) convenes annually, bringing together member states to define and strengthen collective action in response to the escalating threat of climate change. Given the transnational and systemic nature of climate crisis, various international treaties and protocols have been developed to translate scientific consensus into binding commitments and operational measures. For such agreements to acquire legal validity and enforceability, they should undergo formal ratification by the signatory states.

Among these instruments, the Kyoto Protocol remains historically significant as the first treaty to impose legally binding emission reduction targets on industrialized nations (Kim et al., 2020). Its adoption marked a decisive shift from voluntary pledges to enforceable obligations, setting a precedent for subsequent frameworks such as the Paris Agreement (2015), which broadened participation to include all countries and emphasized Nationally Determined Contributions (NDCs) as the core mechanism of global mitigation strategy. Despite these concerted efforts, the global climate system is currently undergoing an unprecedented transformation, characterized by marked increases in surface temperatures, rising sea levels, and intensified extreme weather events. These processes are generating severe and multifaceted impacts across ecosystems, public health, food security, and the world economy, particularly when evaluated from a long-term, intergenerational perspective.

A persistent challenge remains the systematic underestimation of the magnitude and immediacy of climate risks. This tendency toward delayed response and insufficient mitigation represents not only a failure of scientific communication and policy foresight, but also a critical obstacle to ensuring global resilience and sustainable development. Undoubtedly, aggressive and sustained reductions in greenhouse gas emissions will represent the decisive factor in mitigating the magnitude of future climate change (Sarkodie et al., 2020). Nevertheless, owing to the long atmospheric residence time of greenhouse gases, scientific projections indicate that global temperatures may require an additional 20 to 30 years to stabilize, even after emissions reach their peak and begin to decline.

This temporal lag underscores a crucial reality: the climate system's inertia makes delayed action exponentially more costly and less effective. It is, therefore, imperative to act promptly and decisively, pursuing a strategically accelerated transition toward a low-carbon economy. This transformation should be underpinned by the expansive deployment of renewable energy technologies, the systematic decarbonization of production processes, and the mainstreaming of sustainable practices across all sectors of human activity. Only through such an integrated and proactive approach can the international community realistically aspire to limit global warming and safeguard the ecological equilibrium essential to human well-being (Salvati & Carlucci, 2017).

Climate change is profoundly intertwined with socioeconomic dynamics and therefore demands a multidimensional analytical framework. It is not only shaped by social and economic structures but also exerts powerful reciprocal effects on environmental stability, social systems, and national economies. This bidirectional relationship underscores the necessity of adopting integrative approaches capable of addressing both the structural causes and the cascading consequences of climate change. From a geographical standpoint, the use of multiple spatial scales proves indispensable for capturing the complex interconnections among territories and their environmental, social, and political contexts. Such a perspective reveals that climate change cannot be treated as a purely environmental phenomenon; it is instead deeply embedded within broader socioeconomic and political processes that define contemporary globalization.

Economists have long emphasized this interdependence, particularly by examining how the behavior of economic agents, consumers, and firms alike drives greenhouse gas emissions (Papadis & Tsatsaronis, 2020). Consumption patterns, production systems, and technological trajectories, all decisively influence global emission trends. In parallel, the design and implementation of environmental policies and economic strategies emerge as key instruments of mitigation, enabling societies to realign economic incentives with ecological sustainability. Effective climate governance thus requires a comprehensive institutional framework that reconciles growth objectives with the imperative to reduce emissions (Miller et al., 2023).

Empirical and quantitative analyses form the cornerstone of modern economic approaches to climate change. By employing advanced statistical and econometric methodologies, scholars are able to measure the socioeconomic impacts of global warming while simultaneously evaluating the effectiveness of policy interventions and adaptive strategies. Such evidence-based approaches not only quantify the costs of inaction but also provide policymakers with robust decision-support tools (Depledge, 2022). For this reason, all economic measures should rest on rigorous and transparent empirical foundations, ensuring both the credibility of public action and the accountability of climate policy.

In economic theory, climate change is widely conceptualized as a global negative externality, often referred to as the "greatest market failure" in history. Unlike localized externalities, global environmental impacts transcend national borders, generating systemic repercussions for the entire planet. The effects manifested through environmental degradation, economic disruption, and heightened social vulnerability constitute a complex, interdependent risk network that no single nation can effectively confront in isolation (e.g., Salvati & Ferrara, 2013). Consequently, coordinated multilevel governance, encompassing local, national, and international cooperation, becomes indispensable. Only through shared responsibility and strategic global frameworks can the international community mitigate these threats and build a sustainable future.

Human activity—particularly over the last two centuries—has been the main driver of the observed rise in global mean temperatures, a trend projected to exceed

1.5 °C or even 2 °C within the twenty-first century (Adak et al., 2023). The progressive warming of climate systems is empirically evidenced by a multitude of interconnected phenomena: rising sea levels, ocean acidification, glacier retreat, and altered precipitation regimes, which collectively intensify floods, droughts, and extreme weather events. These impacts amplify the risk of irreversible environmental thresholds being crossed, while simultaneously imposing substantial economic and social costs. Among the most critical consequences are the loss of biodiversity, the depletion of water resources, and escalating threats to food security (Semieniuk et al., 2021).

These adverse effects are distributed unevenly across societies. Vulnerable and marginalized communities (i.e., those with limited economic, institutional, and infrastructural capacity) bear a disproportionate share of the burden. Hence, climate change functions not only as an environmental challenge but also as a driver of social inequality, magnifying existing disparities within and between nations. Addressing this multidimensional crisis requires coordinated action at all governance levels. The European Union, for instance, has undertaken ambitious climate policies aimed at reducing greenhouse gas emissions and advancing the transition toward a low-carbon economy. Nonetheless, persistent tensions among member states particularly regarding burden-sharing and the perceived economic costs of restrictive measures highlight the need for governance that is both adaptive and equitable, accommodating the heterogeneous capabilities and socioeconomic contexts of different countries (Salvati & Zitti, 2011).

Ultimately, the success of climate action depends not only on institutional design but also on collective behavioral transformation. Raising public awareness and fostering eco-sustainable practices among citizens and enterprises are indispensable steps toward aligning individual behavior with shared global objectives. Only through such integrated, transparent, and participatory efforts can the international community safeguard planetary systems ensuring not merely the survival of present generations, but also the realization of a sustainable and just future that preserves both environmental integrity and human well-being.

Landscape and Biodiversity

Biodiversity constitutes a foundational pillar of ecological systems, encompassing the full range of variation among living organisms, the genetic information that underpins them, and the ecosystems within which they exist and interact (Cavender-Bares et al., 2022). The term (basically, an abbreviation of biological diversity) was first formalized and popularized in 1986, marking a conceptual turning point in environmental science and conservation policy. Historically, variations within genetic systems and the emergence of new species were primarily governed by natural processes and large-scale geological phenomena. However, in recent decades, human activity has become the predominant agent of ecological disturbance, exerting unprecedented pressure on the planet's biological systems. Despite widespread recognition of biodiversity as essential to the functioning and resilience of living systems, anthropogenic impacts now constitute the main driver of ecological imbalance, threatening the existence of countless species and undermining the natural equilibrium that sustains life (Garole et al., 2025).

The ongoing loss of biodiversity (see Table 4.2 for recent data on global decline) represents a global concern, as human societies derive (direct and indirect) benefits from the stability, productivity, and regulatory services of diverse ecosystems. These include essential functions such as pollination, soil formation, water purification, nutrient cycling, and climate regulation. The erosion of biodiversity therefore poses systemic risks to human well-being and to the long-term sustainability of socioeconomic systems (Sharma & Birman, 2024).

The process of anthropization (namely, the transformation of natural environments through human presence and activity) has accelerated the selective suppression and extinction of various species while artificially favoring others. This asymmetric alteration fundamentally disrupts the integrity of ecological networks and the overall performance of ecosystem services. The resulting decline in functional diversity weakens ecosystems' adaptive capacity, rendering them more vulnerable to both natural and anthropogenic shocks. Mitigating the rapid and continuing loss of living organisms requires a comprehensive and interdisciplinary strategy aimed at restoring the balance between human development and the planet's ecological limits. This entails significant reductions in the consumption of natural resources and energy, alongside the enforcement of robust environmental legislation capable of reshaping current production and consumption models. Above all, it calls for an urgent (global) commitment to transform prevailing lifestyles and

Table 4.2 Selected processes of biodiversity loss (percent rates). Source: Authors' elaboration of recent empirical literature

Economic Process	All species ($n = 1207$)	Vertebrates ($n = 329$)	Invertebrates ($n = 155$)	Plants ($n = 723$)
Agriculture	38%	40%	57%	33%
Cattle grazing	22%	17%	10%	33%
Oil, gas extraction	11%	12%	31%	11%
Cutting wood	12%	16%	25%	7%
Infrastructure	17%	16%	12%	20%
Road	15%	15%	10%	17%
Military activities	4%	2%	1%	5%
Recreational activities	27%	16%	19%	33%
Off-road vehicles	13%	6%	12%	16%
Water networks	30%	47%	66%	15%
Water flow barriers	17%	28%	54%	5%
Pollution	20%	27%	66%	7%
Land conversion	35%	30%	42%	36%
Alteration of fire cycle	14%	5%	6%	20%

economic paradigms, ensuring that future development proceeds within the planet's biophysical boundaries (Hariram et al., 2023). Timely, decisive, and collective action is therefore indispensable to safeguard biodiversity and to secure a viable future for humanity.

The erosion of biodiversity is not merely an ecological concern; it constitutes a profound socioeconomic and ethical challenge, capable of exacerbating existing inequalities and intensifying the marginalization of the most vulnerable sectors within society (Shukla et al., 2021). These groups, often directly dependent on ecosystem services for their subsistence, bear a disproportionate burden when natural systems deteriorate. In managed landscapes, the irreversible loss of genetic information pools further diminishes ecosystem functionality, reducing the adaptive capacity of both natural and agricultural systems to respond to environmental fluctuations and external shocks (O'Connor et al., 2020). The utility of biodiversity, therefore, should be evaluated within a framework that considers the trade-off between costs and benefits of its conservation versus the costs associated with its depletion. While biodiversity conservation often entails substantial upfront financial commitments and may require moderating short-term economic expansion, it generates significant long-term returns, enhanced ecosystem productivity, improved human health and well-being, and expanded opportunities for scientific innovation and bioprospecting. Conversely, the unchecked erosion of biodiversity imposes severe and frequently irreversible costs, including the collapse of critical ecosystem services and heightened vulnerability to climate change (Ekka et al., 2023).

To make sound policy decisions, this trade-off should be quantitatively and objectively assessed, ensuring that management strategies are informed by both immediate and long-term ecological, social, and economic considerations. Achieving this equilibrium between environmental protection, landscape stewardship, and the rational use of natural resources represents the cornerstone of sustainable development and intergenerational equity (Mamabolo et al., 2020). In this context, the Cost–Benefit Analysis (CBA) emerges as an essential methodological instrument (Mishan & Quah, 2020). Cost–Benefit Analysis (CBA) systematically compares the total expected costs and benefits associated with a given decision, project, or policy, determining whether aggregate benefits exceed aggregate costs. As such, it serves as a powerful analytical mechanism for evaluating the feasibility, efficiency, and sustainability of public or private initiatives. Through this framework, economic actors and policymakers can assess the rationality of environmental investments, ensuring that the anticipated social and ecological gains justify the economic outlays (Salvati et al., 2015).

Given the constraints on global resources, the adoption of CBA in environmental policy formulation plays a crucial role in preventing inefficiencies and ensuring optimal resource allocation consistent with the objectives of sustainable development. Nevertheless, analytical rigor alone is insufficient: effective sustainability transitions demand robust political will and behavioral change (e.g., Salvati et al., 2016). In this paradigm shift, human activities should no longer dictate the limits of natural systems; rather, the finite nature of the planet's resources should dictate human behavior and development strategies.

The increasing recognition of biodiversity's economic potential has also led to a complex and often controversial process of commodification whereby elements of biological diversity are transformed into marketable goods through the creation of new property rights, particularly Intellectual Property Rights (IPR) embedded within international trade regimes, notably those of the World Trade Organization (WTO). While these mechanisms aim to stimulate innovation and equitable resource sharing, they often facilitate the privatization of genetic resources and living organisms, effectively transforming them into corporate monopolies through patenting and licensing practices (Matsushita et al., 2015). This commodification, however, stands in direct tension with the very nature of biodiversity as a common good. Common goods—resources managed and utilized collectively for the benefit of all—embody the principles of shared stewardship and collective well-being.

The challenge for sustainable governance lies in reconciling market-based incentives for conservation and innovation with the ethical imperative of maintaining biodiversity as an essential planetary common, ensuring its accessibility and benefits for present and future generations. Scholars have long sought to establish methodologies capable of capturing the economic value of landscape components and ecosystem services, yet this task remains fraught with conceptual and practical limitations. As a matter of fact, biodiversity comprises innumerable species and ecosystems linked through intricate ecological relationships. Capturing this dynamic complexity demands a multidisciplinary approach that transcends reductionist (economic) quantification. Assigning monetary value to biodiversity risks promoting its commercialization and exploitation, potentially undermining conservation efforts and deepening ecological degradation (Salvati et al., 2013). The perceived worth of biodiversity is culturally contingent and context-dependent, varying according to social values. This subjectivity makes the pursuit of a universal, objective valuation highly problematic.

Given these constraints, society should move beyond the narrow framework of "valuation" to practical "valorization"—a process that recognizes and enhances the intrinsic and instrumental importance of biological variety. This shift is essential not only for maintaining ecological equilibrium, but also for countering the unequal dynamics of globalization, which often jeopardize environmental stability and human well-being. The degradation of ecosystems through unsustainable (global) economic models underscores the need to reaffirm biodiversity as an indispensable common asset, rather than a tradable commodity (O'Connor et al., 2020).

Landscape and Social Well-being

In a comprehensive definition of landscape, it is essential to transcend purely physical attributes and incorporate the perception's dimension as experienced by the local community. Only through this interpretative lens can the full value that individuals and societies assign to a landscape be understood—a value that in turn determines its perceived quality (Dong et al., 2022). A circular relationship exists between humanity and the surrounding environment: landscape perception shapes human behavior such as resource use, spatial management, and aesthetic appreciation while human actions modify and reconstruct the landscape. Humanity and landscape thus form an integrated system in which social relations and collective perceptions become key determinants of quality of life (Salvati et al., 2017). Crucially, the quality of human life and the quality of landscapes evolve in parallel, each reinforcing the other. This systemic vision is supported by World Health Organization (WHO), which defines health not merely as the absence of disease but as a state of complete physical, mental, and social well-being.

Within this framework, social relationships and environmental contexts constitute essential determinants of health. Strong, positive social bonds are recognized as key to both physical and psychological well-being, generating direct and measurable benefits for individuals and communities. Landscape quality thus operates not only as a backdrop to human activity but as an active determinant of social and individual well-being. To fully understand this relationship, it is necessary to move beyond qualitative sociological interpretations and rigorously explore well-being from an economic perspective. This brings into focus the field of well-being economics (or "welfare economics"), a specialized branch of economic theory that studies the determinants of human welfare and societal prosperity (Arrow, 2024). Well-being economics rests on the premise that human happiness and progress cannot be captured solely through indicators of material wealth or increases in Gross Domestic Product (GDP). Instead, it recognizes that well-being is shaped by a broad set of intangible and non-monetary factors such as education, healthcare quality, social justice, and the strength of social capital (Ruiz et al., 2011).

The discipline's primary objective is to maximize collective welfare, not individual profit. Within this framework, a key analytical distinction is made between individual well-being, focused on subjective satisfaction and personal fulfillment, and social well-being (or welfare), which assesses objective living conditions, opportunities, and distributive justice across the entire community. This distinction reinforces a multidisciplinary understanding of development: a high quality of life grounded in a healthy landscape, strong social cohesion, and equitable resource distribution offers a more comprehensive and sustainable measure of success than economic output alone.

From a societal perspective, the central problem of welfare economics concerns the optimal allocation of resources among alternative uses, identifying the most efficient and equitable way to employ the resources available to any economic system. This inquiry belongs to normative analysis, which evaluates how changes in economic variables or policy decisions affect welfare outcomes. In this context, the notion of "production functions" is fundamental since it expresses the maximum quantity of goods (or services) that can be produced from a given set of inputs. The point at which no further improvement in one person's welfare can be achieved without diminishing another's welfare is known as the Pareto-optimal situation (Wang et al., 2022).

According to the first theorem of welfare economics, if markets operate under conditions of perfect competition and achieve mutual equilibrium, the resulting allocation of goods and services will be Pareto optimal (Blaug, 2007). Under these conditions, no reallocation of resources can improve one individual's welfare without reducing another's. This theoretical equilibrium aligns with Walrasian thinking, in which prices adjust endogenously to balance supply and demand without external intervention (Misaki, 2023). Walras's law of General Equilibrium further supports this assumption hypothesizing that if equilibrium exists in $n\text{–}1$ markets, it necessarily extends to the n-th market (Walker, 2003).

Notably, such efficiency does not require the conscious pursuit of collective welfare by individuals; rather, it emerges from the decentralized actions of self-interested agents—an insight that echoes Adam Smith's "invisible hand" (Aspromourgos, 2020). However, for equilibrium to exist, several key conditions should be met: utility functions should be continuous and concave, reflecting diminishing marginal utility, and production functions should exhibit decreasing returns to scale, whereby proportional increases in inputs yield less than proportional increases in output (Steeds, 2022). When even a single condition of perfect competition is violated, the Second-Best Theorem becomes relevant. It demonstrates that, if one optimality condition fails, maintaining the others does not necessarily enhance welfare. In fact, social welfare can be maximized by adjusting other parameters. Thus, in real-world economies, lower-order optima are often the attainable goal, which rarely guarantee equity—a critical concern for policymakers.

The second theorem of welfare economics (Tresch, 2022) extends this logic, affirming that a Pareto-optimal allocation can be achieved in competitive equilibrium provided that the initial distribution of resources is equitable. This condition necessitates state intervention, as only public institutions can ensure a fair distribution of initial endowments without distorting market efficiency. Accordingly, a dual structure emerges that includes fairness through central redistribution. At the same time, the market maintains efficiency through decentralized price mechanisms. Nevertheless, applying this theorem in practice is constrained by information asymmetries, resource limitations, and the presence of public goods and externalities—factors that undermine the assumptions of both welfare theorems.

Although idealized, these theorems remain essential for guiding policies that seek to reconcile efficiency, equity, and sustainability. Their principles are especially relevant in the management of environmental and cultural resources which, like all economic goods, should be allocated to maximize net social benefits. Within this framework, minimizing waste and optimizing the aggregate welfare of society become the foundation for evaluating socioeconomic (landscape) values. This analytical approach enhances the potential of sectors such as sustainable tourism and cultural economy, while supporting the broader pursuit of sustainable development meeting present needs without compromising the well-being of future generations.

Toward a "Systemic" Target of Landscape Conservation?

The need of landscape conservation has never been more urgent, particularly given the prolonged periods in which such concerns were underestimated (or neglected) within dominant developmental models (Quaranta et al., 2020). As previously discussed, landscapes constitute a crucial element in shaping the psychophysical well-being of populations, directly enhancing quality of life while simultaneously underpinning the preservation of biodiversity and the protection of (vital) natural and ecological habitats (Foltête et al., 2021). Moving from economic evaluation to (local and regional) governance, it is essential to highlight the adoption of the European Landscape Convention (ELC). This landmark international treaty, signed in Florence (Italy) under the auspices of the Council of Europe in 2000, represents a pivotal moment in the evolution of landscape policy and theory (Pătru-Stupariu & Nita, 2022). The Council of Europe formally defined the landscape as

> *an essential component of people's surroundings, an expression of the diversity of their shared cultural and natural heritage, and a foundation of their identity* (Déjeant-Pons, 2021).

This definition moves beyond a traditional "environmentalist" framework, situating the landscape as both a cultural and natural construct that embodies diversity, identity, and the interdependence between people and place. The ELC thus requires that landscape protection, management, and planning be integrated into all relevant public policies, recognizing the landscape as a determinant of human existence, cultural continuity, and territorial sustainability. The analytical work leading to ELC underscored the dual nature of landscape transformations, which generates both positive and negative externalities. This acknowledgment creates the basis for a new generation of policies that incorporate innovation and technology as indispensable tools for sustainable territorial management and environmental protection. The Convention further emphasizes the central role of public institutions, which bear the responsibility of implementing targeted, coordinated measures to safeguard both the environmental integrity and the cultural–historical identity of landscapes (Salvati et al., 2013).

Despite its holistic scope, the landscape remains a complex analytical category, representing a true "place of interdisciplinary convergence" where physical, social, cultural, and economic factors interact. Such complexity makes its governance particularly challenging. One persistent issue, revealed during ELC implementation, is the tendency of some stakeholders to prioritize only "high-quality" landscapes—especially those of exceptional natural beauty or aesthetic value—while neglecting ordinary (or everyday) landscapes. This selective focus risks undermining the Convention's foundational principle, namely that all landscapes, regardless of their perceived value, contribute to the identity and well-being of their inhabitants (Imbrenda et al., 2021).

In contrast to selective approaches, the European Landscape Convention promotes an inclusive and systemic vision, demanding consideration of the landscape in its totality. This comprehensive approach has posed new challenges for sustainable resource management, particularly concerning UNESCO World Heritage Sites, which should be identified, enhanced, and preserved based on their intrinsic cultural and natural value (Cave & Negussie, 2017). Modernity further complicates this task by introducing new and complex risks arising from rapid technological and infrastructural transformations projects that though innovative, may prove ecologically or culturally disruptive.

Nevertheless, the current distribution of competences and procedures within the national administrative systems in terms of territorial governance remains fragmented and insufficiently coordinated. Addressing this deficiency necessitates a profound reorganization, aimed at ensuring the comprehensive and integrated protection of landscapes. It should recognizes the landscape and all its constituent elements as integral components of the national cultural heritage, thereby reinforcing the interdependence between environmental and cultural protection. This provision acknowledges their essential role in enriching the national territory and safeguarding local cultural traditions and social practices (e.g., Coluzzi et al., 2022).

In the formulation of contemporary landscape management policies, an apparent dichotomy often arises: "Should the landscape be preserved, or should development be promoted?" This dichotomy, however, is conceptually misleading. In modern spatial planning, protection and development should be viewed as complementary objectives following parallel and mutually reinforcing trajectories (Axelsson et al., 2011). Sustainable landscape planning requires maintaining a dynamic equilibrium between these two goals, ensuring that economic and infrastructural progress occurs in harmony with the protection of ecological, historical, and cultural values (e.g.,, Salvati, 2014). Such a balance can only be achieved within a coherent and analytically informed planning framework grounded in a deep understanding of landscape's complexity and multifunctional characters. This approach ensures the coexistence of sustainable development and environmental conservation, thereby securing long-term social well-being, cultural continuity, and ecological resilience.

References

Adak, S., Mandal, N., Mukhopadhyay, A., Maity, P. P., & Sen, S. (2023). Current state and prediction of future global climate change and variability in terms of CO_2 levels and temperature. In *Enhancing resilience of dryland agriculture under changing climate: Interdisciplinary and convergence approaches* (pp. 15–43). Springer Nature.

Araújo, S. O., Peres, R. S., Barata, J., Lidon, F., & Ramalho, J. C. (2021). Characterising the agriculture 4.0 landscape—Emerging trends, challenges and opportunities. *Agronomy, 11*(4), 667.

Arrow, K. J. (2024). An extension of the basic theorems on classical welfare economics. In *The foundations of price theory* (Vol. 5, pp. 253–288). Routledge.

Aspromourgos, T. (2020). Invisible hand. In *The new Palgrave dictionary of economics* (pp. 1–14). Palgrave Macmillan.

Athari, S. A., Alola, U. V., & Alola, A. A. (2023). A global perspective of the role of domestic economic, financial and political risks in inbound tourism. *International Journal of Emerging Markets, 18*(10), 4191–4213.

Axelsson, R., Angelstam, P., Elbakidze, M., Stryamets, N., & Johansson, K. E. (2011). Sustainable development and sustainability: Landscape approach as a practical interpretation of principles and implementation concepts. *Journal of Landscape Ecology, 4*(3), 11–20.

Blaug, M. (2007). The fundamental theorems of modern welfare economics, historically contemplated. *History of Political Economy, 39*(2), 1–10.

Buchadas, A. (2025). *Understanding deforestation frontiers in the world's tropical dry woodlands* (doctoral dissertation), Mathematisch-Naturwissenschaftliche Fakultät, Berlin.

Burkaltseva, D., Tsohla, S., Yanovskaya, A., Guk, O., Zharova, A., & Polskaya, S. (2020). Analysis of the functioning of the tourist and recreational industry in conditions of restrictions. *Journal of Environmental Management & Tourism, 11*(8), 1902–1914.

Canton, H. (2021). World Tourism Organization—UNWTO. In *The Europa directory of international organizations 2021* (pp. 393–397). Routledge.

Cave, C., & Negussie, E. (2017). *World Heritage conservation: The World Heritage Convention, linking culture and nature for sustainable development.* Routledge.

Cavender-Bares, J., Schneider, F. D., Santos, M. J., Armstrong, A., Carnaval, A., Dahlin, K. M., & Wilson, A. M. (2022). Integrating remote sensing with ecology and evolution to advance biodiversity conservation. *Nature Ecology & Evolution, 6*(5), 506–519.

Cheng, L., & Yuan, Y. (2021). Intellectual property tools in safeguarding intangible cultural heritage: A Chinese perspective. *International Journal for the Semiotics of Law, 34*(3), 893–906.

Coluzzi, R., Bianchini, L., Egidi, G., Cudlin, P., Imbrenda, V., Salvati, L., & Lanfredi, M. (2022). Density matters? Settlement expansion and land degradation in Peri-urban and rural districts of Italy. *Environmental Impact Assessment Review, 92*, 106703.

Dangler, J. (2021). Otherness, identities, and cultures in contact. In *The Routledge Hispanic studies companion to medieval Iberia* (pp. 52–66). Routledge.

Déjeant-Pons, M. (2021). The implementation of the European Landscape Convention 2000–2020. In F. M & M. P. Namer (Eds.), *La Convenzione europea del paesaggio vent'anni dopo (2000–2020): Ricezione, criticità, prospettive.* Edizioni Ca' Foscari.

Depledge, J. (2022). The "top-down" Kyoto Protocol? Exploring caricature and misrepresentation in literature on global climate change governance. *International Environmental Agreements: Politics, Law and Economics, 22*(4), 673–692.

Dong, J., Jiang, H., Gu, T., Liu, Y., & Peng, J. (2022). Sustainable landscape pattern: A landscape approach to serving spatial planning. *Landscape Ecology, 37*(1), 31–42.

Duro, J. A., Perez-Laborda, A., Turrion-Prats, J., & Fernández-Fernández, M. (2021). COVID-19 and tourism vulnerability. *Tourism Management Perspectives, 38*, 100819.

Ekka, P., Patra, S., Upreti, M., Kumar, G., Kumar, A., & Saikia, P. (2023). Land degradation and its impacts on biodiversity and ecosystem services. In *Land and environmental management through forestry* (pp. 77–101). Springer Nature.

Estrada-Carmona, N., Sánchez, A. C., Remans, R., & Jones, S. K. (2022). Complex agricultural landscapes host more biodiversity than simple ones: A global meta-analysis. *Proceedings of the National Academy of Sciences, 119*(38), e22033851.

Ezeudu, T., & Chukwudubem, E. K. (2024). The impact of cultural globalization on sustainable urban development in Nigeria. *Journal of Public Administration and Government, 6*(1), 16–34.

Fang, W. T. (2020). *Tourism in emerging economies.* Springer Nature.

Fedyk, W., Sołtysik, M., Bagińska, J., Ziemba, M., Kołodziej, M., & Borzyszkowski, J. (2022). How did the COVID-19 pandemic affect functional relationships in activities between members in a tourism organization? A case study of regional tourism organizations in Poland. *Sustainability, 14*(19), 12671.

Foltête, J. C., Vuidel, G., Savary, P., Clauzel, C., Sahraoui, Y., Girardet, X., & Bourgeois, M. (2021). Graphab: An application for modeling and managing ecological habitat networks. *Software Impacts, 8*, 100065.

Garole, K., Reddy, K. K., & Lippert, K. (2025). Preserving nature's balance in the intelligent systems era: Exploring ecological systems and biodiversity. In *Integration of AI, quantum computing, and semiconductor technology* (pp. 367–390). IGI Global.

Güldner, D., Larsen, L., & Cunfer, G. (2021). Soil fertility transitions in the context of industrialization, 1750–2000. *Social Science History, 45*(4), 785–811.

Hariram, N. P., Mekha, K. B., Suganthan, V., & Sudhakar, K. (2023). Sustainalism: An integrated socio-economic-environmental model to address sustainable development and sustainability. *Sustainability, 15*(13), 10682.

Heritage, U. I. C., & Rii, P. (2020). Convention for the safeguarding of the intangible cultural heritage. In *Proceedings of the eleventh annual coordination meeting of category 2 Centres, New York* (Vol. 2).

Hristov, J., Clough, Y., Sahlin, U., Smith, H. G., Stjernman, M., Olsson, O., & Brady, M. V. (2020). Impacts of the EU'S common agricultural policy "greening" reform on agricultural development, biodiversity, and ecosystem services. *Applied Economic Perspectives and Policy, 42*(4), 716–731.

Imbrenda, V., Quaranta, G., Salvia, R., Egidi, G., Salvati, L., Prokopovà, M., Coluzzi, R., & Lanfredi, M. (2021). Land degradation and metropolitan expansion in a peri-urban environment. *Geomatics, Natural Hazards and Risk, 12*(1), 1797–1818.

Jackson, G. A. (2024). The first agro-industrial revolution. In *New era—New urgency: The case for repurposing education* (p. 41). Springer.

Javaid, M., Haleem, A., Singh, R. P., & Suman, R. (2022). Enhancing smart farming through the applications of agriculture 4.0 technologies. *International Journal of Intelligent Networks, 3*, 150–164.

Jeon, C. Y., & Yang, H. W. (2021). The structural changes of a local tourism network: Comparison of before and after COVID-19. *Current Issues in Tourism, 24*(23), 3324–3338.

Kala, N. (2021). Chronological review of the adoption of the World Tourism Organization basic international regulatory acts on tourism. *Journal of Legal, Ethical & Regulatory Issues, 24*(1), 2–10.

Kaplan, D., & Holloway, S. (2024). *Urban geography*. Wiley.

Kim, Y., Tanaka, K., & Matsuoka, S. (2020). Environmental and economic effectiveness of the Kyoto Protocol. *PLoS One, 15*(7), e0236299.

Knoll, M., & Reith, R. (2020). Mobilities, migration and demography. In *Concepts of urban-environmental history* (pp. 203–220). Springer.

Knowles, R. D., Ferbrache, F., & Nikitas, A. (2020). Transport's historical, contemporary and future role in shaping urban development: Re-evaluating transit-oriented development. *Cities, 99*, 102607.

Kudryavtseva, V. A. (2020). Modern city: Areas of modernization of the urban environment. In *IOP conference series: Materials science and engineering* (Vol. 880(1), p. 012064). IOP.

Langlais, A. (2023). The new common agricultural policy: Reflecting an agro-ecological transition. *Review of Agricultural, Food and Environmental Studies, 104*(1), 51–66.

Lazaro Ortiz, S., & Jimenez de Madariaga, C. (2022). The UNESCO convention for the safeguarding of the intangible cultural Heritage: A critical analysis. *International Journal of Cultural Policy, 28*(3), 327–341.

Llorca-Rodríguez, C. M., García-Fernández, R. M., & Casas-Jurado, A. C. (2020). Domestic versus inbound tourism in poverty reduction: Evidence from panel data. *Current Issues in Tourism, 23*(2), 197–216.

Mamabolo, E., Makwela, M. M., & Tsilo, T. J. (2020). Achieving sustainability and biodiversity conservation in agriculture: Importance, challenges and prospects. *European Journal of Sustainable Development, 9*(3), 616.

Matsushita, M., Schoenbaum, T. J., Mavroidis, P. C., & Hahn, M. (2015). *The World Trade Organization: Law, practice, and policy*. Oxford University Press.

Meehl, G. A., Covey, C., Delworth, T., Latif, M., McAvaney, B., Mitchell, J. F., & Taylor, K. E. (2007). The WCRP CMIP3 multimodel dataset: A new era in climate change research. *Bulletin of the American Meteorological Society, 88*(9), 1383–1394.

Meyer, M. C., Gliganic, L. A., May, J. H., Merchel, S., Rugel, G., Schlütz, F., & Krainer, K. (2020). Landscape dynamics and human–environment interactions in the northern foothills of Cho Oyu

and Mount Everest (Southern Tibet) during the late Pleistocene and Holocene. *Quaternary Science Reviews, 229*, 106127.
Miller, A., Atakhanov, A., Guliyev, M., Azizov, T., & Huseynova, K. (2023). The economic effect of the measures provided for by the Kyoto protocol by region. *Scientific Horizons, 26*(4), 136–145.
Misaki, K. (2023). *Léon Walras's economic thought: The general equilibrium theory in historical perspective*. Routledge.
Mishan, E. J., & Quah, E. (2020). *Cost-benefit analysis*. Routledge.
Mom, G. (2020). Globalizing automobilism: Exuberance and the emergence of layered mobility, 1900–1980.
O'Connor, B., Bojinski, S., Röösli, C., & Schaepman, M. E. (2020). Monitoring global changes in biodiversity and climate essential as ecological crisis intensifies. *Ecological Informatics, 55*, 101033.
O'Sullivan, J. N. (2020). The social and environmental influences of population growth rate and demographic pressure deserve greater attention in ecological economics. *Ecological Economics, 172*, 106648.
Papadis, E., & Tsatsaronis, G. (2020). Challenges in the decarbonization of the energy sector. *Energy, 205*, 118025.
Patel, A. (2023). Traditional knowledge, sustainability, and international intellectual property law: Biopiracy in patent-intensive industries. In *Sustainable boardrooms: Democratising governance and technology for society and economy* (pp. 115–140). Springer Nature.
Pătru-Stupariu, I., & Nita, A. (2022). Impacts of the European landscape convention on interdisciplinary and transdisciplinary research. *Landscape Ecology, 37*(5), 1211–1225.
Pillai, S. D., Goldfarb, B., & Kirsch, D. A. (2020). The origins of firm strategy: Learning by economic experimentation and strategic pivots in the early automobile industry. *Strategic Management Journal, 41*(3), 369–399.
Quaranta, G., Salvia, R., Salvati, L., Paola, V. D., Coluzzi, R., Imbrenda, V., & Simoniello, T. (2020). Long-term impacts of grazing management on land degradation in a rural community of southern Italy: Depopulation matters. *Land Degradation & Development, 31*(16), 2379–2394.
Romero-Rodríguez, L. M., Civila, S., & Aguaded, I. (2021). Otherness as a form of intersubjective social exclusion: Conceptual discussion from the current communicative scenario. *Journal of Information, Communication and Ethics in Society, 19*(1), 20–37.
Ruiz, V. R. L., Navarro, J. L. A., & Peña, D. N. (2011). Relationship between gross domestic product (GDP) and hidden wealth over the period 2000–2008: An international study. *Electronic Journal of Knowledge Management, 9*(3), 259–270.
Salvati, L. (2014). The spatial pattern of soil sealing along the urban–rural gradient in a Mediterranean region. *Journal of Environmental Planning and Management, 57*(6), 848–861.
Salvati, L., & Carlucci, M. (2013). The impact of Mediterranean land degradation on agricultural income: A short-term scenario. *Land Use Policy, 32*, 302–308.
Salvati, L., & Carlucci, M. (2014). Zero net land degradation in Italy: The role of socioeconomic and agro-forest factors. *Journal of Environmental Management, 145*, 299–306.
Salvati, L., & Carlucci, M. (2015). Towards sustainability in agro-forest systems? Grazing intensity, soil degradation and the socioeconomic profile of rural communities in Italy. *Ecological Economics, 112*, 1–13.
Salvati, L., & Carlucci, M. (2016). The way towards land consumption: Soil sealing and polycentric development in Barcelona. *Urban Studies, 53*(2), 418–440.
Salvati, L., & Carlucci, M. (2017). Urban growth, population, and recession: Unveiling multiple spatial patterns of demographic indicators in a Mediterranean city. *Population, Space and Place, 23*(8), e2079.
Salvati, L., & Ferrara, C. (2013). Do changes in vegetation quality precede urban sprawl? *Area, 45*(3), 365–375.

Salvati, L., & Zitti, M. (2011). Economic growth vs. land quality: A multidimensional approach in Italy. *Journal of Environmental Planning and Management, 54*(6), 733–748.

Salvati, L., Tombolini, I., Perini, L., & Ferrara, A. (2013). Landscape changes and environmental quality: The evolution of land vulnerability and potential resilience to degradation in Italy. *Regional Environmental Change, 13*, 1223–1233.

Salvati, L., Biasi, R., Carlucci, M., & Ferrara, A. (2015). Forest transition and urban growth: Exploring latent dynamics (1936–2006) in Rome, Italy, using geographically weighted regression and implications for coastal forest conservation. *Rendiconti Lincei, 26*, 577–585.

Salvati, L., Venanzoni, G., Serra, P., & Carlucci, M. (2016). Scattered or polycentric? Untangling urban growth in three southern European metropolitan regions through exploratory spatial data analysis. *The Annals of Regional Science, 57*, 1–29.

Salvati, L., Carlucci, M., & Venanzoni, G. (2017). Recession, resilience, local labour markets: Wealthier is better? *Letters in Spatial and Resource Sciences, 10*(2), 177–204.

Sarkodie, S. A., Adams, S., & Leirvik, T. (2020). Foreign direct investment and renewable energy in climate change mitigation: Does governance matter? *Journal of Cleaner Production, 263*, 121262.

Seboka, A. D., Ewunie, G. A., Morken, J., Feng, L., & Adaramola, M. S. (2024). Potentials and prospects of solid biowaste resources for biofuel production in Ethiopia: A systematic review of the evidence. *Biomass Conversion and Biorefinery, 14*(24), 30929–30960.

Semieniuk, G., Taylor, L., Rezai, A., & Foley, D. K. (2021). Plausible energy demand patterns in a growing global economy with climate policy. *Nature Climate Change, 11*(4), 313–318.

Sevilla-Buitrago, Á. (2022). *Against the commons: A radical history of urban planning*. University of Minnesota Press.

Sharma, I., & Birman, S. (2024). Biodiversity loss, ecosystem services, and their role in promoting sustainable health. In *The climate-health-sustainability nexus: Understanding the interconnected impact on populations and the environment* (pp. 163–188). Springer Nature.

Shukla, K., Shukla, S., Upadhyay, D., Singh, V., Mishra, A., & Jindal, T. (2021). Socio-economic assessment of climate change impact on biodiversity and ecosystem services. In *Climate change and the microbiome: Sustenance of the ecosphere* (pp. 661–694). Springer.

Solomon, B. D. (2023). Intergovernmental panel on climate change (IPCC). In *Dictionary of ecological economics* (p. 302). Edward Elgar.

Sorensen, A. (2020). Urbanization and developmental pathways: Critical junctures of urban transition. In *Handbook of megacities and megacity-regions* (pp. 47–64). Edward Elgar.

Steeds, L. (2022). The social ecology of Adam Smith: Reconsidering the intellectual foundations of political economy. *New Political Economy, 27*(1), 132–145.

Svoray, T. (2022). Soil erosion: The general problem. In *A geoinformatics approach to water erosion: Soil loss and beyond* (pp. 1–38). Springer Nature.

Tresch, R. W. (2022). *Public finance: A normative theory*. Academic Press.

Trusova, N. V., Kyrylov, Y. Y., Hranovska, V. H., Prystemskyi, O. S., Krykunova, V. M., & Sakun, A. Z. (2020). The imperatives of the development of the tourist services market in spatial polarization of the regional tourist system. *Geojournal of Tourism and Geosites, 29*(2), 565–582.

Vărzaru, A. A., Bocean, C. G., & Cazacu, M. (2021). Rethinking tourism industry in pandemic COVID-19 period. *Sustainability, 13*(12), 6956.

Veal, A. J. (2025). Human rights, travel and tourism. In *Leisure and human rights* (pp. 203–237). Springer Nature.

Vicente-Vicente, J. L., Quintas-Soriano, C., & López-Rodríguez, M. D. (2022). A transformative (r)evolution of the research on agriculture through fostering human–nature connectedness: A special issue editorial. *Agriculture, 12*(4), 522.

Walker, D. A. (2003). Early general equilibrium economics: Walras, Pareto, and Cassel. In *A companion to the history of economic thought* (pp. 278–293). Routledge.

Wang, H., Xie, Z., Pu, L., Ren, Z., Zhang, Y., & Tan, Z. (2022). Energy management strategy of hybrid energy storage based on Pareto optimality. *Applied Energy, 327*, 120095.

Wells, J. C., & Stock, J. T. (2020). Life history transitions at the origins of agriculture: A model for understanding how niche construction impacts human growth, demography and health. *Frontiers in Endocrinology, 11*, 325.

Wiktor-Mach, D. (2020). What role for culture in the age of sustainable development? UNESCO's advocacy in the 2030 agenda negotiations. *International Journal of Cultural Policy, 26*(3), 312–327.

Zhai, Z., Martínez, J. F., Beltran, V., & Martínez, N. L. (2020). Decision support systems for agriculture 4.0: Survey and challenges. *Computers and Electronics in Agriculture, 170*, 105256.

Chapter 5
Economic Growth and Landscape Change: Unraveling the Development Path of Local Systems

Abstract Global warming, biodiversity loss, soil degradation, air and water pollution, deforestation, and stratospheric ozone depletion demonstrates that humanity has effectively become a geological force, capable of irreversibly altering the equilibrium of ecosystems. Sustainable development underscores the necessity of viewing sustainability as a systemic process rather than a static equilibrium, demanding continuous adaptation and innovation. The analysis addresses the inherently complex and non-linear interactions between socioecological systems, landscapes, and economic structures, highlighting the decisive role of externalities and the monetary valuation of ecosystem services. Such relationships demonstrate that environmental degradation and resource depletion are not only ecological issues but also structural failures of market and policy systems unable to internalize the true costs of environmental impacts. The perspective adopted in this work emphasizes the urgent need for policies and strategies grounded in intergenerational responsibility and the transition toward innovative economic paradigms, including innovation economy. These approaches represent essential pathways for achieving a sustainable future with technological progress, cultural evolution, and economic development.

Keywords Anthropocene · Ecosystem services · Socioecological complexity · Environmental externalities · Circular economy

Introduction

Humankind has emerged as the primary driver of regional transformations, landscape dynamics, and the alteration of ecosystem functionality across multiple spatial and temporal scales. Over the course of 4.5 billion years, terrestrial ecosystems have undergone both gradual and abrupt transitions, some of which can be classified as catastrophic events. The most recent geological epoch, the Quaternary, has been characterized by alternating glacial and interglacial phases as well as the early manifestations of climate variability (Anderson et al., 2013). Increasingly, contemporary evidence attributes such dynamics, at least in part, to anthropogenic forces,

L. Salvati et al., *Landscape Economics*, Advances in Spatial Science,
https://doi.org/10.1007/978-3-032-25263-0_5

with global warming being structurally and functionally associated with rising atmospheric concentrations of carbon dioxide and other greenhouse gases (Salvati et al., 2014).

Combustion processes have intensified the emission of aerosols, while ozone depletion in the troposphere has been directly associated with the release of ChloroFluoroCarbons (CFCs) from industrial activities (Wilson et al., 2007). Geological studies document the onset of a new anthropogenic epoch: in the past two centuries, atmospheric carbon dioxide levels have increased by approximately 30%, and methane concentrations have doubled (Kabir et al., 2023). These trends, unequivocally anthropogenic in origin, substantiate the classification of the present epoch as the Anthropocene—an era defined by the unprecedented capacity of human activity to reshape Earth system processes, with profound and enduring geological implications (Gibbard et al., 2022a, b).

The term *Anthropocene* was introduced and popularized by Paul J. Crutzen, Nobel Laureate in Chemistry, whose pioneering research on atmospheric dynamics fundamentally advanced the understanding of human influence on the Earth system (e.g., Crutzen & Brauch, 2016; Bertomeu Sánchez, 2016;. Benner et al., 2021). A landmark publication by members of the Working Group on the Anthropocene, entitled "The Anthropocene is functionally and stratigraphically distinct from the Holocene" (Lax, 2022), provides compelling evidence that the Anthropocene has already left a distinct geological imprint, differentiating it from all preceding epochs (Waters et al., 2016). Indicators of this transition include geochemical signatures such as radioactive isotopes and the proliferation of novel anthropogenic materials, including aluminum, concrete, and plastics, now recognized as rapidly evolving "technofossils" (Zalasiewicz et al., 2014). Parallel to these markers, human activities such as mining, fossil fuel extraction, deforestation, urbanization, coastal modification, and intensive agriculture have profoundly altered soil composition and sedimentary processes worldwide. Over the past century, global stocks of nitrogen and phosphorus have more than doubled due to fertilizer use, leaving measurable imprints in lake sediments and glacial cores (Anderson et al., 2018).

These dynamics are accompanied by global warming and accelerated rates of species extinction, emphasizing the magnitude of human-driven transformations of the Earth system (Salvati et al., 2017a, b). The onset of the Anthropocene is generally associated with the Industrial Revolution, when mechanization and technological innovation initiated the large-scale exploitation of environmental resources (Brolin & Kander, 2022). Notably, as early as 1873, Italian geologist Antonio Stoppani referred to an "anthropozoic era," describing humanity as a "new telluric force" shaping the planet (Rull, 2021). Similarly, in 1926, Russian geologist and biologist Vladimir Vernadsky (following Kautzleben & Müller, 2014) observed that

> *the surface of the Earth has been transformed beyond recognition, and there is no doubt that greater transformations will take place in the future. We are confronted with a new form of biogenic migration resulting from the activity of human reason.*

While the long-term consequences of human impact remain difficult to predict, the potential to develop technologies and strategies capable of ensuring prosperity for both humanity and other forms of life offers a pathway to re-establishing a more balanced relationship with the Earth system (Lenka, 2020). Among territorial stakeholders, the growing awareness of the interconnections among climate, landscapes, and ecosystems, together with the evolving role of scientific communities, is fostering new opportunities for transformative action toward global sustainability. The notion of sustainable development emerged in the late 1970s, following recognition that industrialized societies were increasingly incompatible with ecological constraints—particularly regarding resource exploitation and the negative externalities generated by environmentally intensive production activities (McNeill, 2004; Springett & Redclift, 2015; Staniškis, 2022). From the late 1980s onward, and especially during the 1990s, it became evident that the economy, society, and environment should not be regarded as distinct domains, but rather as interdependent dimensions within a unified framework oriented toward genuine sustainability (e.g., Salvati et al., 2016).

The most widely cited definition of sustainable development is provided by the Brundtland Report (1987), prepared by the World Commission on Environment and Development, which defines it as a "development that meets the needs of the present without compromising the ability of future generations to meet their own needs" (Barkemeyer et al., 2014). A complementary formulation was proposed in the report entitled "Caring for the Earth: A Strategy for Sustainable Living" (1991), published by the International Union for Conservation of Nature (IUCN), the United Nations Environment Programme (UNEP), and the World Wide Fund for Nature (WWF). This report emphasizes "the improvement of quality of life while remaining within the carrying capacity of the ecosystems that sustain us" (Graymore et al., 2010).

Both definitions frame sustainable development as a dual responsibility: first, to ensure intergenerational equity by preserving opportunities for future generations; and second, to maintain the regenerative capacity of ecosystems (Keong, 2020). These principles underscore the intrinsic interdependence between human well-being and the integrity of natural environments (Visser, 2017). The past may be interpreted as a sequence of experiences often marked by environmental degradation that provide the empirical foundation for reorienting collective strategies toward higher standards of ecological integrity and social well-being (Drastichová, 2022). If the human element is central to the relationship between communities and their surrounding environments, it follows that education plays a fundamental role in advancing sustainability. Education specifically acts as a catalyst for transformation, enabling individuals to recognize their capacity and responsibility to function as agents of change. It also fosters values, behaviors, and lifestyles consistent with sustainability, and equips individuals to make decisions oriented toward the long-term well-being of local communities.

Given that anthropogenic pressures are among the most significant factors shaping climate dynamics and transforming landscapes and biomes, awareness and education should rest on solid scientific, theoretical, and practical foundations (Dar et al., 2020). This calls for the integration of diverse (technical, empirical, and operational) disciplinary approaches to produce actionable insights for continuous environmental monitoring and the formulation of effective conservation policies. Within this framework, the present contribution situates itself at the intersection of multiple domains of knowledge, reinterpreting issues traditionally framed as environmental (or ecological) such as landscapes and ecosystems, through the lens of social sciences, with particular attention to their economic implications (Salvati et al., 2019).

This approach acknowledges that the evolving relationship between humans and the environment unfolds through non-linear spatio–temporal dynamics, demanding interpretive models capable of addressing systemic complexity. While linear models offer analytical simplicity and transparency, the real-world dynamics of socio-ecological systems are inherently non-linear, characterized by limited predictability and path dependence. Consequently, this work assumes that local systems are dynamic entities evolving along non-linear developmental trajectories. By adopting a discursive yet rigorous approach grounded in complexity theory, we define Socio-Ecological Systems (SES), particularly at the local scale, as the elementary analysis' units for understanding the co-evolution of economy and environment (Han et al., 2025). These dynamics can be explored in greater detail with a focus on the dialectical relationship between economic space and landscape.

Here, the landscape is not conceived merely as an aesthetic or static entity but as the continuous and dynamic outcome of human interaction with the natural environment, representing the spatial manifestation of historical interrelations between social structures and ecosystems. From this perspective, the analysis will propose methodologies assigning value to the landscape in both biophysical (ecological) and monetary (economic) terms. On this conceptual foundation, we will introduce the topic of the monetary evaluation of ecosystem services as the essential operational concept that bridges the analytical gap between structure (landscape) and functions (ecological dynamics and economic processes). Accordingly, this study adopts the complexity of economic–environmental relationships as its central methodological principle and interpretive axiom (Han et al., 2023). Such an approach is indispensable for achieving a genuine integration of concepts and disciplinary frameworks originating from diverse scientific fields, in pursuit of a coherent understanding of local socioenvironmental evolution.

Anthropocene Landscapes

The Anthropocene, the geological epoch defined by the unprecedented scale of human activities that fundamentally reshape ecological processes and determine long-term planetary trajectories, is characterized by profound geomorphological transformations (Gibbard et al., 2022a, b). These transformations are primarily associated with the accelerated modification of natural processes such as erosion and sediment transport now overwhelmingly mediated by anthropogenic drivers. The intensification of land use through industrial-scale agriculture, rapid urbanization, and the extensive conversion of natural ecosystems has disrupted established hydrological regimes and soil dynamics (Zalasiewicz et al., 2021).

Collectively, these processes contribute to the formation of a pervasive anthropogenic stratum within the Earth's sedimentary archive, providing material evidence that contemporary economic activity no longer influences—but actively drives—the planet's geological evolution (James, 2013). In this context, the socioeconomic metabolism of modern societies constitutes the primary engine of such transformations (Haberl et al., 2023). This metabolism is characterized by systemic overexploitation of natural resources, disproportionate waste generation, and the exponential growth of emissions. Despite advances in technology and environmental policy, these pressures continue to intensify, sustained by self-reinforcing feedback mechanisms that amplify systemic environmental stress. The resulting global effects (e.g., biodiversity loss and climate change) represent only the most visible manifestations of the Anthropocene's impact. The scope of environmental science, however, extends far beyond these phenomena and requires precise conceptual framing. According to the European Environment Agency (EEA), ecological impact is defined as the alteration of the biological, physical, and chemical properties of the environment when it creates a risk (or potential risk) to human health and the well-being of any living species, the deterioration of material goods, or damage to the recreational values of the environment (Salvia et al., 2018).

Within this framework, various forms of pollution constitute tangible expressions of environmental alteration. The most pervasive among these processes is atmospheric pollution, defined as any modification in air composition resulting from the emission of substances harmful to organisms and ecosystems. However, environmental contamination extends across multiple domains including water, soil, and acoustic environments, reflecting an underlying disequilibrium in the socioecological system. Air pollution remains one of the most critical threats to planetary health (McKinney, 2019). The atmosphere acts as a vast reservoir for thousands of pollutants originating from both stationary sources (industrial facilities, domestic heating systems) and mobile sources (vehicular traffic). The resulting impact on human and ecological health depends primarily on the toxicity, concentration, and exposure duration of these substances.

A second (major) dimension of degradation is water pollution, which arises when human activities introduce mutagenic or carcinogenic compounds into hydrological systems—for instance, through the discharge of industrial effluents or agricultural runoff. Continuous monitoring of drinking water quality is essential, as many of these contaminants disrupt endocrine functions and damage genetic material, leading to severe reproductive and developmental disorders (e.g., Serra et al., 2016). The analysis of aquatic systems is inherently complex, given their spatial and temporal variability in chemical and biological composition. Effective monitoring should therefore account for the type of water body (surface, groundwater, coastal), flow regime, and seasonal dynamics. One of the most destructive consequences of water pollution is eutrophication, a process driven by nutrient over-enrichment, particularly nitrogen and phosphorus inputs (Tiwari & Pal, 2022). This excess fosters algal blooms whose subsequent decomposition depletes dissolved oxygen, generating extensive hypoxic "dead zones" in marine and lacustrine environments and causing drastic losses in biodiversity and productivity.

Soil pollution represents another interconnected form of degradation, primarily resulting from the prolonged use of agrochemicals, improper waste disposal, and contamination from mining or industrial processes (Devlin & Brodie, 2023). Because soil functions as a filtering and storage medium, it often accumulates toxic substances that ultimately infiltrate aquifers, linking soil degradation to groundwater contamination (Koul & Taak, 2018). The loss of soil fertility and productive capacity also undermines the biogeochemical balance essential for ecosystem stability. Among the most visible and ecologically consequential processes of land degradation is deforestation (Kumar et al., 2022).

Beyond the physical removal of trees, deforestation entails the conversion of forest land to alternative uses (agricultural monocultures, urban settlements, or extensive grazing areas) triggering cascading effects such as erosion, altered microclimates, and accelerated biodiversity decline (Salvati et al., 2015). Although forests are theoretically renewable resources, this definition lacks practical validity when deforestation and soil degradation rates systematically exceed natural or managed regeneration capacities. Forest ecosystems, which host over half of terrestrial biodiversity, play a crucial role in carbon sequestration and climate regulation, while providing more than five thousand commercially significant products, including timber and cellulose (Shukla et al., 2021).

Closely related is the phenomenon of soil degradation, a cumulative process leading to the partial or complete loss of arable topsoil and the long-term decline of its productive capacity (Wang et al., 2023). This process, often irreversible on human timescales, disrupts vital biogeochemical cycles and can precipitate desertification, particularly in ecologically fragile regions such as arid and semi-arid zones (Laflen & Roose, 2020). According to the United Nations Convention to Combat Desertification (UNCCD), desertification is a

> *land degradation in arid, semi-arid, and dry sub-humid areas resulting from various factors, including climatic variations and human activities* (Akhtar-Schuster et al., 2022).

Mitigating this phenomenon requires large-scale reforestation and afforestation programs aimed at stabilizing soils, restoring hydrological cycles, and reinforcing the resilience of vulnerable socioecological systems (Islam et al., 2025). Another critical dimension of environmental change is stratospheric ozone depletion, first observed in the 1970s. The ozone layer, situated within the stratosphere, plays a protective role by filtering ultraviolet (UV) radiation from solar emissions. Excessive UV exposure causes irreversible damage to living organisms and disrupts key ecological processes. The recognition of the so-called *ozone hole* in the mid-1980s mobilized an extraordinary international response, culminating in the adoption of the Montreal Protocol (1989) on Substances that Deplete the Ozone Layer (Mäder et al., 2010).

This landmark treaty mandated the global phase-out of CFCs and other ozone-depleting substances. The Protocol's success hailed by former United Nations Secretary-General Kofi Annan as "an example of exceptional international cooperation, probably the most successful agreement between nations"—remains a cornerstone of global environmental governance. The most severe ozone thinning occurs over Antarctica during the austral spring, where it alters not only the biosphere's exposure to UV radiation but also atmospheric dynamics, including wind circulation and oceanic current systems (Bornman et al., 2019). These perturbations further illustrate the systemic interconnectedness between atmospheric chemistry and surface climate processes (Zerefos et al., 2023).

Environmental Protection, Landscape Quality, and Sustainable Development

In the contemporary context of accelerating climate change and intensifying anthropogenic pressure on ecosystems, environmental protection assumes a fundamental and non-negotiable role in securing the planet's long-term sustainability. Given this centrality, the term *environment* has acquired multiple, and at times divergent, definitions. A precise conceptual delineation is essential, as a significant semantic gap persists between the term's colloquial use and its rigorous scientific and academic application. By adopting a more integrative understanding, the environment can be broadly defined as everything that surrounds us and is capable, in one way or another, of influencing our lives, both through material needs and, at least in part, spiritual ones (Salvati et al., 2011).

From this geographical perspective, the environment encompasses not only its ecological dimension but also the profound imprint of anthropization—the cumulative transformation of natural systems through human activity. Within this analytical framework, the Life Cycle Assessment (LCA) methodology emerges as a pivotal tool (Bishop et al., 2021). LCA provides a comprehensive and rigorous framework for evaluating the environmental impacts associated with a product, process, service, or even a specific landscape throughout its entire life cycle from the extraction of raw materials ("cradle") to production, use, and eventual disposal or recycling ("grave"). Its primary objective is to quantify and assess potential burdens based on established sustainability indicators such as energy consumption, GreenHouse Gas (GHG) emissions, waste generation, and water use (Batool et al., 2024). Through this systematic approach, abstract environmental concerns are transformed into measurable, actionable metrics, thereby enhancing both transparency and decision-making capacity.

In contemporary discourse, the concepts of sustainability and sustainable development are often used interchangeably, yet their meanings are not identical. "Sustainability" refers fundamentally to the capacity of a system—economic, ecological, or sociocultural—to persist over time by withstanding external shocks without depleting its essential resources. Within the domain of productive activities, economic sustainability may be defined as the ability of an enterprise (or system) to maintain stable income and prosperity in the long term, without compromising the welfare or productive potential available to future generations (Sevigné-Itoiz et al., 2021). This systemic concept of sustainability intersects with the dynamic and action-oriented process of sustainable development. Although widely popularized by the 1992 "Rio Earth Summit" (Grubb et al., 2019), the concept was first formally articulated in the 1987 Brundtland Report ("Our Common Future"), which remains the cornerstone of global sustainability policy. The World Commission on Environment and Development (Borowy, 2013) defines sustainable development as a

> *development that meets the needs of the present without compromising the ability of future generations to meet their own needs.*

This definition foregrounds the principle of intergenerational equity, situating sustainable development as a process through which the equilibrium implied by sustainability can be operationally achieved (Spijkers, 2018). Sustainable development thus represents a model integrating three mutually interdependent dimensions—environmental, economic, and social—whose balanced coexistence constitutes the so-called *three-sphere* model of sustainability. Within this triadic structure, particular emphasis is placed on social sustainability, understood as a society's ability to meet the needs of its members in the present, without jeopardizing the capacity of future generations to do likewise (Tepecik Diş & Karimnia, 2021). Central to this dimension is the concept of social justice, which entails the equitable distribution of resources, rights, and opportunities (Fedotova et al., 2023). Such equity is indispensable for reducing inequalities and for fostering active community participation in decision-making processes that directly affect societal and environmental well-being.

Closely related to social sustainability is the notion of capital legacies, which underscores the ethical responsibility of the present generation to preserve and transmit to future generations a stock of natural, social, and economic capital that is at least equivalent both qualitatively and quantitatively to what it currently possesses (Pierce et al., 2016). These cumulative stocks represent the material and immaterial basis of collective well-being. Scholars commonly distinguish between "weak" and "strong" sustainability. "Weak sustainability" conceptualizes natural capital as one component of total capital, assuming a high degree of substitutability between natural resources and human-made capital (Ekins, 2014).

Within this framework, market mechanisms and technological innovation are deemed sufficient to sustain aggregate capital over time, provided that the benefits of resource exploitation are reinvested in new forms of productive or social capital. "Strong sustainability," by contrast, affirms the non-substitutable and intrinsic value of natural capital, recognizing the existence of critical ecological functions indispensable for the survival of both human and non-human systems (Cisternas et al., 2024). Consequently, it rejects the assumption of perfect substitutability between natural and artificial capital.

This distinction has become increasingly salient in contemporary environmental economics and governance, where strong sustainability is widely regarded as a robust theoretical and ethical framework. It integrates both intergenerational and intragenerational justice as essential components of social welfare, while also foregrounding the concept of "ecological footprint," namely a quantitative measure of the biophysical resources required to sustain current patterns of consumption and production. In practice, however, economic imperatives often take precedence over ecological considerations, leading to a short-term, individualistic orientation that undermines the regenerative capacity of natural systems. This erosion of environmental sensitivity compromises the provision of essential ecosystem services from climate regulation to water purification and soil fertility maintenance (Ekka et al., 2023). The central challenge, therefore, lies in harmonizing the objectives of efficiency, effectiveness, and sustainability.

Here, "efficiency" denotes the ability to achieve predefined goals, while "effectiveness" refers to the capacity of utilizing resources with minimal waste. Their integration within a sustainability framework yields a more comprehensive understanding of the evolutionary dynamics of economic systems. Such alignment can be strengthened through the adoption of advanced environmental accounting tools, which enable the systematic analysis of official datasets to assess environmental performance. These instruments assist enterprises in evaluating their ecological impacts, formulating more sustainable strategies, optimizing resource use, and reducing operational costs. Ultimately, the integration of sustainability principles into economic decision-making fosters responsible resource management, enhances corporate accountability, and reinforces the interdependence between environmental integrity and societal well-being (Izah et al., 2025).

Economic Growth and the Landscape

Economic development can be accurately defined as a transformative process aimed at improving the economic conditions of a community, region, or country (Storper, 2005). Its fundamental objective is to facilitate sustainable economic growth through strategic interventions that expand job opportunities and raise per capita income. In the contemporary global context, these strategies increasingly promote investment in innovation and technological advancement, while simultaneously addressing the persistent challenges of poverty alleviation and the reduction of economic and social inequality. Within this complex process, the inclusion of cultural factors and the prioritization of educational attainment are indispensable (Salvati et al., 2017a, b). They function as non-material yet essential inputs for sustained progress. Without robust investment in these areas, economic development risks stagnation and eventual decline, constrained by diminishing returns on physical and natural resources (e.g., Kosmas et al., 2015).

Culture should therefore be recognized not as a peripheral byproduct but as a strategic asset and a superior good within economic development (Dudin et al., 2013). For superior goods, consumption increases disproportionately with rising real income. Similarly, the utility of cultural capital exhibits cumulative properties: the more individuals engage with and consume this asset through education, knowledge acquisition, and shared social values. This dynamic generates positive externalities that reinforce long-term, knowledge-based economic prosperity. The traditional dichotomy between economic structures and non-material societal factors is increasingly obsolete; economy and culture are fundamentally interdependent (e.g., Salvati, 2013). Cultural activities and creative industries are not merely consumption sectors but powerful engines of measurable economic growth, particularly through employment generation and value creation. As dynamic, space- and time-sensitive activities, these sectors drive innovation, attract talent, and enhance regional attractiveness, thereby contributing directly to sustainable economic progress.

The trajectory of global economic development has been profoundly shaped by successive Industrial Revolutions, each introducing radical technological paradigms from mechanical energy and steam power to digitalization and biotechnology (Rusko & Kosonen, 2023). These revolutions have expanded development opportunities worldwide but also accentuated structural disparities, leaving many regions struggling with economic marginalization and underdevelopment (Candeias, 2023). The combined influence of technological innovation, globalization, and the drastic reduction in transportation and communication costs has fundamentally restructured the International Division of Labor, IDL (Freyssenet, 2012).

This restructuring has led to new forms of global specialization, that include peripheral specialization (concentration in low-cost, high-volume manufacturing or primary production of basic goods, typically associated with limited value-added processes) and core specialization (concentration in high-technology sectors, knowledge-intensive services, and innovation-driven industries, which capture a disproportionate share of global value creation). This dichotomy underscores the strategic importance of technological and cultural capital, confirming that access to (and investment in) these forms of capital are critical determinants of a nation's competitive position within the global economy.

The evolving structure of the IDL and the uneven spatial distribution of development directly fuel migratory flows (Charnock & Starosta, 2016). Individuals move (voluntarily or involuntarily) in search of better employment and living conditions (Erokhin et al., 2023). This dynamic is bidirectional: the global demand for low-cost labor, often supplied by migrant populations, also shapes the geographical distribution of production processes, leading to complex transnational labor circuits. The level of economic development in a society is typically assessed through various key performance indicators, for example, per-capita income—a metric that captures the total earnings from wages, salaries, and other returns, reflecting an individual's capacity to meet personal and familial needs. While invaluable, income alone is an incomplete measure of well-being. Structural determinants such as education level and unemployment rate should also be considered, as these correlate with long-term economic performance. Empirical evidence consistently shows that higher educational attainment corresponds with lower unemployment rates, confirming education's dual role as both a productivity enhancer and a stabilizer against structural labor market fluctuations (Sayer & Morgan, 2018).

Economic development also produces externalities indirect effects, positive or negative, experienced by third parties not directly involved in the economic transaction (Lecocq et al., 2023). Positive externalities generate social benefits, such as job creation or community-wide increases in income; negative externalities in turn impose social or environmental costs, notably through pollution, habitat destruction, or resource depletion. Recognizing these externalities is essential to internalizing their costs and benefits within policy frameworks, ensuring that development strategies promote holistic and sustainable prosperity rather than short-term economic gains. Externalities are inherently market failures, as they lie outside the price system and thus evade accurate valuation (Macháč et al., 2021). Economic theory classifies them as technological externalities (real effects on production or utility, such as pollution or innovation spillovers), or pecuniary externalities (effects mediated through price changes, such as shifts in wages or rents). Both can be further differentiated by their scope.

Atmospheric externalities include indivisible effects impacting many agents simultaneously within a defined area. Specific externalities instead represent localized effects limited to a single actor or entity. A classic atmospheric (positive) externality is found in knowledge spillovers within industrial clusters (Xu et al., 2022), where firms benefit collectively from shared expertise, tacit knowledge, and specialized labor. Similarly, investments in pollution abatement technologies (Bag & Pretorius, 2022) generate environmental and health benefits extending well beyond the adopting firm, illustrating how localized initiatives can yield global positive outcomes. By contrast, specific externalities such as the diffusion of unpatented innovations or localized efficiency gains tend to produce benefits confined to a smaller spatial (or organizational) scale, though still economically and socially meaningful. Pecuniary externalities often arise within specialized industrial districts, where co-location enables firms to achieve lower input costs through economies of scale, specialized services, and aggregated demand.

Landscape quality indirectly shapes these advantages: an unpolluted, aesthetically valuable environment enhances local productivity and competitiveness by improving access to quality resources, skilled labor, and reduced transaction costs. Specific pecuniary externalities such as the opening of a successful retail enterprise can revitalize neighborhoods, attract foot traffic, and increase commercial activity, providing uncompensated benefits to surrounding businesses. Therefore, managing both positive and negative externalities through well-designed regulatory and fiscal mechanisms is fundamental to ensuring a development trajectory that is equitable, sustainable, and durable.

Since the 1990s, economic research has sought empirical frameworks to describe the intricate relationship between development and environmental quality (Lau et al., 2025). One of the most influential is the Environmental Kuznets Curve (EKC) (Gill et al., 2018). The EKC hypothesizes an inverted U-shaped relationship between income and environmental degradation: pollution initially rises with industrialization but eventually declines after a critical income threshold is reached, as societies invest in cleaner technologies and stricter environmental standards (Leal & Marques, 2022). While the EKC provides a compelling narrative of potential 'green growth', its empirical robustness remains contested (Fig. 5.1). The model's optimistic implications that prosperity naturally leads to environmental improvement are undermined by inconsistent data, particularly in developing countries lacking reliable statistics (Amaefule et al., 2022). For this reason, high-quality, disaggregated data are indispensable for robust analysis in environment–development research. Only through precise, empirically grounded models can policymakers effectively balance economic growth with environmental sustainability, ensuring that progress today does not compromise the ecological foundations of tomorrow (e.g., Kairis et al., 2014).

Finally, it is essential to highlight the diverse theoretical approaches to economic growth that inform both policy and academic debate. Endogenous Growth Theory (EGT) is one of these foundational frameworks (Akcigit & Ates, 2021). This model posits that economic development is generated internally and is primarily sustained through strategic investments in physical, human, and technological capital. By incentivizing Research and Development (R&D) and fostering innovation, the model seeks to continuously increase productivity and counteract the limitations of diminishing returns that characterize simpler growth models (Lamonica et al., 2020). Alternatively, Modernization Theory views economic development as the result of a linear series of processes, key among them being the sociocultural and political transformations that historically follow urbanization and industrialization. This perspective emphasizes how changes in social structure and political institutions generate shifts in the economic and geopolitical fields, ultimately determining a plurality of territorial outcomes.

Regardless of the model adopted, the exploitation of natural resources has historically been the empirical basis for economic growth. However, reliance on this exploitation mechanism, despite its capacity to generate rapid growth, fundamentally risks becoming unsustainable in the long term. A more recent and critical approach to achieving sustainable economic development is based on the concept of

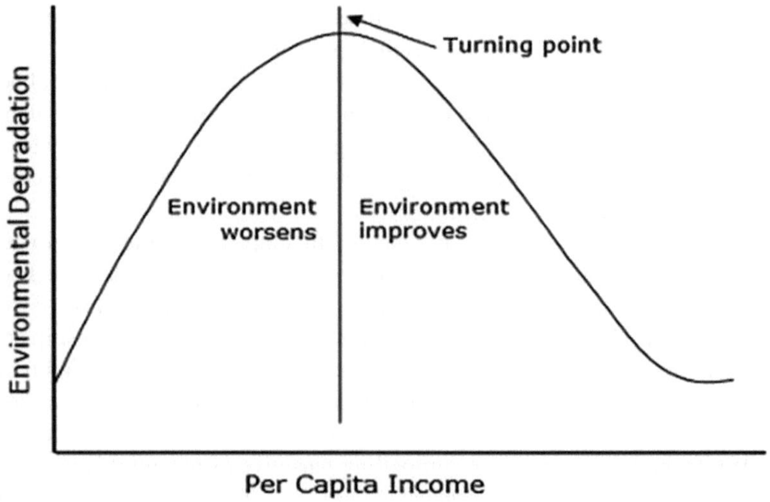

Fig. 5.1 Schematic illustration of the Environmental Kuznets Curve (EKC). (Source: Authors' elaboration of recent literature in environmental economics)

Circular Economy (Corvellec et al., 2022). This model promotes recycling, remanufacturing, and reuse of goods after their initial use, aiming to minimize waste and reduce the total material throughput of the economy. By keeping resources in use for as long as possible, the circular economy contributes to the creation of a truly sustainable economic system that drastically reduces environmental impact (George et al., 2015). The approach not only facilitates the reduction of production costs, but also catalyzes the underlying process of innovation. In this sense, Innovation Economy is equally vital. This framework prioritizes the creation of new and original products, alongside the development of new markets and manufacturing processes. By channeling strategic investments into scientific and technological research, the innovation economy manages to increase the competitiveness of firms, thereby promoting economic growth and socioecological resilience.

This framework has gained particular attention over the last two decades, reshaping international technological competitiveness and redefining the complex relationship between technology, employment, and overall economic expansion (Ekins et al., 2020). To conclude, economic development is fundamentally a complex, multilayered process involving a wide array of intertwined strategies associated with the perennial goals of job opportunity creation and the increase of individual income levels (Robert & Yoguel, 2011). Any comprehensive analysis of this growth process should holistically account for structural factors (such as education and unemployment rates) but, most importantly, should internalize the critical notion of externalities. These unpriced effects significantly impact the quality of development, generating either systemic benefits or uncompensated damages. Ultimately, integrating these diverse factors is essential for guaranteeing genuinely sustainable economic development that lasts over time and genuinely enhances population wealth and well-being.

References

Akcigit, U., & Ates, S. T. (2021). Ten facts on declining business dynamism and lessons from endogenous growth theory. *American Economic Journal: Macroeconomics, 13*(1), 257–298.

Akhtar-Schuster, M., Stringer, L. C., Metternicht, G., Barger, N. N., Chotte, J. L., & Kust, G. (2022). Assessing the impact of science in the implementation of the United Nations convention to combat desertification. *Land, 11*(4), 568.

Amaefule, C., Kalu, I., Udeorah, S., Ibeabuchi, I., Adeola, S., & Ebelebe, L. (2022). Transposed second-generation environmental Kuznets curve, changing climate patterns, and selected development indicators. *European Journal of Sustainable Development Research, 6*(4), 1–11.

Anderson, D. E., Goudie, A., & Parker, A. (2013). *Global environments through the quaternary: Exploring environmental change*. Oxford University Press.

Anderson, N. J., Curtis, C. J., Whiteford, E. J., Jones, V. J., McGowan, S., Simpson, G. L., & Kaiser, J. (2018). Regional variability in the atmospheric nitrogen deposition signal and its transfer to the sediment record in Greenland lakes. *Limnology and Oceanography, 63*(5), 2250–2265.

Bag, S., & Pretorius, J. H. C. (2022). Relationships between Industry 4.0, sustainable manufacturing and circular economy: Proposal of a research framework. *International Journal of Organizational Analysis, 30*(4), 864–898.

Barkemeyer, R., Holt, D., Preuss, L., & Tsang, S. (2014). What happened to the "development" in sustainable development? Business guidelines two decades after Brundtland. *Sustainable Development, 22*(1), 15–32.

Batool, F., Kurniawan, T. A., Mohyuddin, A., Othman, M. H. D., Aziz, F., Al-Hazmi, H. E., & Anouzla, A. (2024). Environmental impacts of food waste management technologies: A critical review of life cycle assessment (LCA) studies. *Trends in Food Science & Technology, 143*, 104287.

Benner, S., Lax, G., Crutzen, P. J., Pöschl, U., Lelieveld, J., & Brauch, H. G. (Eds.). (2021). *Paul J. Crutzen and the Anthropocene: A new epoch in earth's history*. Springer.

Bertomeu Sánchez, J. R. (2016). Paul J. Crutzen: A pioneer on atmospheric chemistry and climate change in the Anthropocene. *Ambix, 63*(4), 359–360.

Bishop, G., Styles, D., & Lens, P. N. (2021). Environmental performance comparison of bioplastics and petrochemical plastics: A review of life cycle assessment (LCA) methodological decisions. *Resources, Conservation and Recycling, 168*, 105451.

Bornman, J. F., Barnes, P. W., Robson, T. M., Robinson, S. A., Jansen, M. A., Ballaré, C. L., & Flint, S. D. (2019). Linkages between stratospheric ozone, UV radiation and climate change and their implications for terrestrial ecosystems. *Photochemical & Photobiological Sciences, 18*(3), 681–716.

Borowy, I. (2013). *Defining sustainable development for our common future: A history of the world commission on environment and development (Brundtland commission)*. Routledge.

Brolin, J., & Kander, A. (2022). Global trade in the Anthropocene: A review of trends and direction of environmental factor flows during the Great Acceleration. *The Anthropocene Review, 9*(1), 71–110.

Candeias, M. (2023). International division of labour. In *Historical-critical dictionary of Marxism* (pp. 449–461). Brill.

Charnock, G., & Starosta, G. (2016). Introduction: The new international division of labour and the critique of political economy today. In *The new international division of labour: Global transformation and uneven development* (pp. 1–22). Palgrave Macmillan.

Cisternas, M. E., Faucheu, J., Troussier, N., & Laforest, V. (2024). Implementing strong sustainability in a design process. *Cleaner Environmental Systems, 15*, 100224.

Corvellec, H., Stowell, A. F., & Johansson, N. (2022). Critiques of the circular economy. *Journal of Industrial Ecology, 26*(2), 421–443.

Crutzen, P. J., & Brauch, H. G. (2016). *Paul J. Crutzen: A pioneer on atmospheric chemistry and climate change in the Anthropocene* (p. 248). Springer.

Dar, J. A., Subashree, K., Bhat, N. A., Sundarapandian, S., Xu, M., Saikia, P., & Khan, M. L. (2020). Role of major forest biomes in climate change mitigation: An eco-biological perspective. In *Socio-economic and eco-biological dimensions in resource use and conservation: Strategies for sustainability* (pp. 483–526). Springer.

Devlin, M., & Brodie, J. (2023). Nutrients and eutrophication. In *Marine pollution: Monitoring, management and mitigation* (pp. 75–100). Springer.

Drastichová, M. (2022). Sustainable development and sustainable science. Where we came from, where we are now and where we are heading? Part I: The history of the concept. *Problemy Ekorozwoju, 17*(2), 7–18.

Dudin, M., Lyasnikov, N., Kuznecov, A., & Fedorova, I. (2013). Innovative transformation and transformational potential of socio-economic systems. *Economic Systems*, 1434–1437.

Ekins, P. (2014). Strong sustainability and critical natural capital. In *Handbook of sustainable development* (pp. 55–71). Edward Elgar.

Ekins, P., Domenech Aparisi, T., Drummond, P., Bleischwitz, R., Hughes, N., & Lotti, L. (2020). *The circular economy: What, why, how and where*. Routledge.

Ekka, P., Patra, S., Upreti, M., Kumar, G., Kumar, A., & Saikia, P. (2023). Land degradation and its impacts on biodiversity and ecosystem services. In *Land and environmental management through forestry* (pp. 77–101). Springer Nature.

Erokhin, V., Tianming, G., & Andrei, J. V. (2023). International migration of labor. In *Contemporary macroeconomics: New global disorder* (pp. 781–816). Springer Nature.

Fedotova, I., Bocharova, N., Rachwał-Mueller, A., Byrski, W., Grabiński, T., & Luty, Z. (2023). Corporate social responsibility as an instrument of sustainable business development: Exploring types and dimensions. *Zeszyty Naukowe Wyższej Szkoły Ekonomii i Informatyki w Krakowie, 19*, 83–110.

Freyssenet, M. (2012). The intellectual division of labour: A stake in the current crisis? In *Keynote presented at the 30th international labour process conference, Stockholm* (pp. 27–29).

George, D. A., Lin, B. C. A., & Chen, Y. (2015). A circular economy model of economic growth. *Environmental Modelling & Software, 73*, 60–63.

Gibbard, P. L., Bauer, A. M., Edgeworth, M., Ruddiman, W. F., Gill, J. L., Merritts, D. J., & Ellis, E. C. (2022a). A practical solution: The Anthropocene is a geological event, not a formal epoch. *Episodes: Journal of International Geoscience, 45*(4), 349–357.

Gibbard, P., Walker, M., Bauer, A., Edgeworth, M., Edwards, L., Ellis, E., & Ruddiman, W. (2022b). The Anthropocene as an event, not an epoch. *Journal of Quaternary Science, 37*(3), 395–399.

Gill, A. R., Viswanathan, K. K., & Hassan, S. (2018). The environmental Kuznets curve (EKC) and the environmental problem of the day. *Renewable and Sustainable Energy Reviews, 81*, 1636–1642.

Graymore, M. L., Sipe, N. G., & Rickson, R. E. (2010). Sustaining human carrying capacity: A tool for regional sustainability assessment. *Ecological Economics, 69*(3), 459–468.

Grubb, M., Koch, M., Thomson, K., Sullivan, F., & Munson, A. (2019). The "Earth Summit" agreements: A guide and assessment. In *An analysis of the Rio '92 UN conference on environment and development*. Routledge.

Haberl, H., Fischer-Kowalski, M., Krausmann, F., & Schmid, M. (2023). Socio-metabolic transitions. In *Introduction to designing environments: Paradigms & approaches* (pp. 71–92). Springer.

Han, Z., Wei, Y., Meng, J., Zou, Y., & Wu, Q. (2023). Integrated water security and coupling of social-ecological system to improve river basin sustainability. *Science of the Total Environment, 905*, 167182.

Han, Z., Wei, Y., & Meng, J. (2025). Systemic and explicit approaches to assessing the impact of natural and human drivers on the co-evolution of social-ecological systems in the Heihe River Basin, China. *Environmental Impact Assessment Review, 112*, 107757.

Islam, W., Zeng, F., Siddiqui, J. A., Zhihao, Z., Du, Y., Zhang, Y., & Khan, K. A. (2025). Combating desertification: Comprehensive strategies, challenges, and future directions for sustainable solutions. *Biological Reviews, a1082*.

Izah, S. C., Ogwu, M. C., & Salimon, S. S. (2025). Principles of sustainable development and environmental management. In *Evaluating environmental processes and technologies* (pp. 29–63). Springer Nature.

James, L. A. (2013). Legacy sediment: Definitions and processes of episodically produced anthropogenic sediment. *Anthropocene, 2*, 16–26.

Kabir, M., Habiba, U. E., Khan, W., Shah, A., Rahim, S., De los Rios-Escalante, P. R., & Shafiq, M. (2023). Climate change due to increasing concentration of carbon dioxide and its impacts on environment in 21st century: A mini review. *Journal of King Saud University—Science, 35*(5), 102693.

Kairis, O., Kosmas, C., Karavitis, C., Ritsema, C., Salvati, L., Acikalin, S., Alcalá, M., Alfama, P., Atlhopheng, J., Barrera, J., Belgacem, A., Solé-Benet, A., Brito, J., Chaker, M., Chanda, R., Coelho, C., Darkoh, M., Diamantis, I., Ermolaeva, O., …, & Ziogas A. (2014). Evaluation and selection of indicators for land degradation and desertification monitoring: types of degradation, causes, and implications for management. *Environmental Management, 54*, 971–982.

Kautzleben, H., & Müller, A. (2014). Vladimir Ivanovich Vernadsky (1863–1945)—From mineral to noosphere. *Journal of Geochemical Exploration, 147*, 4–10.

Keong, C. Y. (2020). *Global environmental sustainability: Case studies and analysis of the United Nations' journey toward sustainable development*. Elsevier.

Kosmas, C., Kairis, O., Karavitis, C., Acikalin, S., Alcalá, M., Alfama, P., Atlhopheng, J., Barrera, J., Belgacem, A., Solé-Benet, A., Brito, J., Chaker, M., Chanda, R., Darkoh, M., Ermolaeva, O., Fassouli, V., Fernandez, F., Gokceoglu, C., Gonzalez, D., … Salvati, L. (2015). An exploratory analysis of land abandonment drivers in areas prone to desertification. *Catena, 128*, 252–261.

Koul, B., & Taak, P. (2018). Soil pollution: Causes and consequences. In *Biotechnological strategies for effective remediation of polluted soils* (pp. 1–37). Springer Nature.

Kumar, R., Kumar, A., & Saikia, P. (2022). Deforestation and forests degradation impacts on the environment. In *Environmental degradation: Challenges and strategies for mitigation* (pp. 19–46). Springer.

Laflen, J. M., & Roose, E. J. (2020). Methodologies for assessment of soil degradation due to water erosion. In *Methods for assessment of soil degradation* (pp. 31–55). Springer.

Lamonica, G. R., Recchioni, M. C., Chelli, F. M., & Salvati, L. (2020). The efficiency of the cross-entropy method when estimating the technical coefficients of input–output tables. *Spatial Economic Analysis, 15*(1), 62–91.

Lau, L. S., Yii, K. J., Ng, C. F., Tan, Y. L., & Yiew, T. H. (2025). Environmental Kuznets curve (EKC) hypothesis: A bibliometric review of the last three decades. *Energy & Environment, 36*(1), 93–131.

Lax, G. (2022). Paul J. Crutzen and the path to the Anthropocene. In J. Paul (Ed.), *Crutzen and the Anthropocene: A new epoch in earth's history* (pp. 1–15). Springer.

Leal, P. H., & Marques, A. C. (2022). The evolution of the environmental Kuznets curve hypothesis assessment: A literature review under a critical analysis perspective. *Heliyon, 8*(11), e11312.

Lecocq, X., Demil, B., & Warnier, V. (2023). Moving business models forward for positive social, environmental and economic outcomes: Managing externalities (and internalities) for sustainability innovation. *Journal of Business Models, 11*(3), 117–132.

Lenka, N. (2020). Philosophical and scientific interaction between Vladimir Vernadsky and Pavel Florensky. *Вестник Санкт-Петербургского университета. Философия и конфликтология, 36*(4), 645–656.

Macháč, J., Trantinová, M., & Zaňková, L. (2021). Externalities in agriculture: How to include their monetary value in decision-making? *International journal of Environmental Science and Technology, 18*(1), 3–20.

Mäder, J. A., Staehelin, J., Peter, T., Brunner, D., Rieder, H. E., & Stahel, W. A. (2010). Evidence for the effectiveness of the Montreal protocol to protect the ozone layer. *Atmospheric Chemistry and Physics, 10*(24), 12161–12171.

McKinney, A. (2019). A planetary health approach to study links between pollution and human health. *Current Pollution Reports, 5*(4), 394–406.

McNeill, D. (2004). The concept of sustainable development. *Development Studies and Political Ecology in a North South Perspective, 5*, 26–46.

Pierce, J. C., Lovrich, N. P., & Budd, W. W. (2016). Social capital, institutional performance, and sustainability in Italy's regions: Still evidence of enduring historical effects? *The Social Science Journal, 53*(3), 271–281.

Robert, V., & Yoguel, G. (2011). The complex dynamics of economic development. In *Handbook on the economic complexity of technological change*. Edward Elgar.

Rull, V. (2021). The Anthropozoic era revisited. *Lethaia, 54*(3), 289–299.

Rusko, R., & Kosonen, M. (2023). Emerging new technologies and industrial revolution. In *Encyclopedia of data science and machine learning* (pp. 2161–2181). IGI Global.

Salvati, L. (2013). Monitoring high-quality soil consumption driven by urban pressure in a growing city (Rome, Italy). *Cities, 31*, 349–356.

Salvati, L., Bajocco, S., Ceccarelli, T., Zitti, M., & Perini, L. (2011). Towards a process-based evaluation of land vulnerability to soil degradation in Italy. *Ecological Indicators, 11*(5), 1216–1227.

Salvati, L., Ranalli, F., & Gitas, I. (2014). Landscape fragmentation and the agro-forest ecosystem along a rural-to-urban gradient: An exploratory study. *International Journal of Sustainable Development and World Ecology, 21*(2), 160–167.

Salvati, L., Kosmas, C., Kairis, O., Karavitis, C., Acikalin, S., Belgacem, A., Solé-Benet, A., Chaker, M., Fassouli, V., Gokceoglu, C., Gungor, H., Hessel, R., Khatteli, H., Kounalaki, A., Laouina, A., Ocakoglu, F., Ouessar, M., Ritsema, C., Sghaier, M., … De Vente, J. (2015). Unveiling soil degradation and desertification risk in the Mediterranean basin: A data mining analysis of the relationships between biophysical and socioeconomic factors in agro-forest landscapes. *Journal of Environmental Planning and Management, 58*(10), 1789–1803.

Salvati, L., Kosmas, C., Kairis, O., Karavitis, C., Acikalin, S., Belgacem, A., Solé-Benet, A., Chaker, M., Fassouli, V., Gokceoglu, C., Gungor, H., Hessel, R., Khatteli, H., Kounalaki, A., Laouina, A., Ocakoglu, F., Ouessar, M., Ritsema, C., Sghaier, M., … Carlucci, M. (2016). Assessing the effectiveness of sustainable land management policies for combating desertification: A data mining approach. *Journal of Environmental Management, 183*, 754–762.

Salvati, L., De Zuliani, E., Sabbi, A., Cancellieri, L., Tufano, M., Caneva, G., & Savo, V. (2017a). Land-cover changes and sustainable development in a rural cultural landscape of Central Italy: Classical trends and counter-intuitive results. *International Journal of Sustainable Development and World Ecology, 24*(1), 27–36.

Salvati, L., Ranalli, F., Carlucci, M., Ippolito, A., Ferrara, A., & Corona, P. (2017b). Forest and the city: A multivariate analysis of peri-urban forest land cover patterns in 283 European metropolitan areas. *Ecological Indicators, 73*, 369–377.

Salvati, L., Serra, P., Bencardino, M., & Carlucci, M. (2019). Re-urbanizing the European city: A multivariate analysis of population dynamics during expansion and recession times. *European Journal of Population, 35*(1), 1–28.

Salvia, R., Serra, P., Zambon, I., Cecchini, M., & Salvati, L. (2018). In-between sprawl and neo-rurality: Sparse settlements and the evolution of socio-demographic local context in a Mediterranean region. *Sustainability, 10*(10), Article 3670.

Sayer, A., & Morgan, K. (2018). High technology industry and the international division of labour: The case of electronics. In *The development of high technology industries* (pp. 10–36). Routledge.

Serra, P., Salvati, L., Queralt, E., Pin, C., Gonzalez, O., & Pons, X. (2016). Estimating water consumption and irrigation requirements in a long-established Mediterranean rural community by remote sensing and field data. *Irrigation and Drainage, 65*(5), 578–588.

Sevigné-Itoiz, E., Mwabonje, O., Panoutsou, C., & Woods, J. (2021). Life cycle assessment (LCA): Informing the development of a sustainable circular bioeconomy? *Philosophical Transactions of the Royal Society A, 379*(2206), 20200352.

Shukla, K., Shukla, S., Upadhyay, D., Singh, V., Mishra, A., & Jindal, T. (2021). Socio-economic assessment of climate change impact on biodiversity and ecosystem services. In *Climate change and the microbiome: Sustenance of the ecosphere* (pp. 661–694). Springer.

Spijkers, O. (2018). Intergenerational equity and the sustainable development goals. *Sustainability, 10*(11), 3836.

Springett, D., & Redclift, M. (2015). Sustainable development: History and evolution of the concept. In *Routledge international handbook of sustainable development* (pp. 3–38). Routledge.

Staniškis, J. K. (2022). Socio-environmental-economic transformations towards sustainable development. In *Transformation of business organization towards sustainability: Systems approach* (pp. 81–165). Springer.

Storper, M. (2005). Society, community, and economic development. *Studies in Comparative International Development, 39*(4), 30–57.

Tepecik Diş, A., & Karimnia, E. (2021). Reframing Kiruna's relocation—Spatial production or a sustainable transformation? *Sustainability, 13*, 3811.

Tiwari, A. K., & Pal, D. B. (2022). Nutrients contamination and eutrophication in the river ecosystem. In *Ecological significance of river ecosystems* (pp. 203–216). Elsevier.

Visser, W. (2017). Our common future ('The Brundtland Report') world commission on environment and development (1987). In *The top 50 sustainability books* (pp. 52–55). Routledge.

Wang, J., Zhen, J., Hu, W., Chen, S., Lizaga, I., Zeraatpisheh, M., & Yang, X. (2023). Remote sensing of soil degradation: Progress and perspective. *International Soil and Water Conservation Research, 11*(3), 429–454.

Waters, C. N., Zalasiewicz, J., Summerhayes, C., Barnosky, A. D., Poirier, C., Gałuszka, A., & Wolfe, A. P. (2016). The Anthropocene is functionally and stratigraphically distinct from the Holocene. *Science, 351*(6269), aad2622.

Wilson, S. R., Solomon, K. R., & Tang, X. (2007). Changes in tropospheric composition and air quality due to stratospheric ozone depletion and climate change. *Photochemical & Photobiological Sciences, 6*(3), 301–310.

Xu, Y., Li, X., Tao, C., & Zhou, X. (2022). Connected knowledge spillovers, technological cluster innovation and efficient industrial structure. *Journal of Innovation & Knowledge, 7*(3), 100195.

Zalasiewicz, J., Williams, M., Waters, C. N., Barnosky, A. D., & Haff, P. (2014). The technofossil record of humans. *The Anthropocene Review, 1*(1), 34–43.

Zalasiewicz, J., Waters, C. N., Ellis, E. C., Head, M. J., Vidas, D., Steffen, W., & Zinke, J. (2021). The Anthropocene: Comparing its meaning in geology (chronostratigraphy) with conceptual approaches arising in other disciplines. *Earth-Science Reviews, 216*, 103593.

Zerefos, C., Fountoulakis, I., Eleftheratos, K., & Kazantzidis, A. (2023). Long-term variability of human health-related solar ultraviolet-B radiation doses from the 1980s to the end of the 21st century. *Physiological Reviews, 103*(3), 1789–1826.

Chapter 6
Evaluating Ecosystem Services and Landscape Quality

Abstract This chapter examines the multifaceted relationship between human societies, economic development, and the environment, emphasizing how anthropogenic pressures have profoundly redefined both ecological dynamics and landscape transformations. Situated within the broader discourse on the Anthropocene, the discussion highlights the extent to which industrialization, urbanization, and intensive resource exploitation have produced unprecedented environmental consequences, including climate change, biodiversity loss, and the widespread degradation of soil and water systems. Particular attention is devoted to social justice and intergenerational equity as prerequisites for achieving long-term socioecological resilience. The work further explores the valuation of ecosystem services and underscores the importance of advanced environmental accounting tools capable of quantifying externalities and guiding organizations toward more sustainable production and management practices.

Keywords Socioecological systems · Environmental externalities · Weak sustainability · Circular economy · Human–environment interactions

Introduction

Evaluating landscapes constitutes a fundamental methodology, as it enables the accurate estimation of both physical and monetary attributes of a given territory (or reference area). Analytical assessment of landscapes represents a cognitive process that integrates natural and anthropogenic dimensions, thereby generating a robust informational foundation for strategic planning and decision-making. Such evaluations are intended not only to preserve and enhance territorial assets but also to ensure their long-term ecological and socioeconomic sustainability. In regional studies, the combined use of quantitative analysis expressed in both physical and economic terms plays a decisive role in determining the intrinsic and functional value of landscapes (Maghsoudi et al., 2024). The significance of a landscape is often underestimated; yet, through rigorous evaluation of its biophysical features

L. Salvati et al., *Landscape Economics*, Advances in Spatial Science,
https://doi.org/10.1007/978-3-032-25263-0_6

and economic potential, its uniqueness and societal importance can be effectively demonstrated. Consequently, landscape evaluation emerges as an essential tool for guiding policies aimed at enhancing and safeguarding territories while respecting both environmental integrity and the well-being of local communities. Given that landscape quality is a key determinant of human welfare, the absence of standardized methodologies for its quantitative assessment represents a critical gap in current research and practice (Smiraglia et al., 2019).

Throughout history, scholars have developed increasingly sophisticated approaches grounded in statistical, spatial, and geographical techniques to analyze landscape quality in relation to integrity, vulnerability, and transformation over time. These methods provide indispensable support for defining Landscape Quality Objectives (LQOs) and for identifying areas requiring conservation, enhancement, or redevelopment. Descriptive statistics can offer a valuable framework for quantitative landscape analysis. However, their application should account for intrinsic limitations and contextual constraints. A primary limitation lies in the restrictive assumptions underlying analytical techniques. Within landscape research, this assumption proves especially problematic, as landscapes are inherently heterogeneous systems characterized by complex interactions, diverse processes, and dynamic transformations. Consequently, the uncritical application of standardized statistical models risks oversimplifying spatial reality and masking significant ecological and socioterritorial variations (Vardopoulos et al., 2023).

Another methodological challenge derives from the necessity of conducting integrated analysis across multiple spatial and temporal scales. In landscape studies, the scales of observation and interpretation often differ substantially from those employed in conventional models, necessitating methodological adaptations to ensure that data are appropriately contextualized and ecologically meaningful (Tombolini et al., 2016). Without such adjustments, results may lack both analytical validity and practical relevance. A further challenge concerns the interpretation of outcomes, as landscapes comprise interdependent physical, biological, and social components that interact in non-linear and context-dependent ways.

This complexity renders the translation of quantitative findings into meaningful insights considerably more difficult than in controlled analytical settings. For these reasons, while statistical approaches remain an invaluable component of landscape evaluation, their effective implementation requires both methodological flexibility and a deep understanding of the specific characteristics of the landscape under investigation (e.g., Simoniello et al., 2022). Ultimately, quantitative analysis should be context-sensitive, combining statistical rigor with ecological awareness and geographical interpretive depth, ensuring that the inherent complexity of socioecological systems can be adequately captured and meaningfully represented.

Geographic Information Systems

From this perspective, a variety of tools can be employed in the quantitative evaluation of landscapes, with Geographic Information Systems (GIS) having a central role (Lü et al., 2019). GIS technology enables the collection, integration, analysis, and visualization of spatial data including maps, satellite imagery, and topographic information, thereby providing a powerful means of understanding both the structure and dynamics of landscapes (Fradelos et al., 2014). More than a purely technical instrument, GIS embodies the systemic paradigm of integrated spatial thinking, combining analytical precision with relational data management in ways that have transformed both research methodologies and professional practices. Over the past three decades, GISs have expanded particularly in the Anglo-Saxon world (e.g., Brookes, 2007), evolving from an academic and research-oriented tool into a system of substantial commercial and institutional importance. Its widespread diffusion is largely attributable to its capacity to integrate heterogeneous datasets and support complex analysis that would otherwise be difficult (or impossible) to perform (e.g., Vinci et al., 2023).

The architecture of GIS rests on several key components. At its core lies the Relational DataBase Management System (RDBMS), which serves as the backbone for storing and managing spatially referenced information (Zlatanova & Stoter, 2006). This database is directly associated with spatial viewers and analytical tools capable of performing advanced queries across well-defined cartographical units. The integration of these components forms what is known as a Geographic Relational DataBase Management System (GRDBMS), enabling sophisticated spatial queries, simulations, and decision-support processes. Thus, GISs transcend the role of simple cartographical tool, functioning instead as a comprehensive analytical environment for territorial evaluation and spatial planning. Conceptually, the definition of GIS rests on three terms: *system*, *information*, and *geographic*. The word "system" underscores the integrative nature of GIS as a framework combining multiple components into a holistic analytical model. The notion of "information" recognizes data as a resource of the modern economy comparable to labor, capital, or equipment in the production of goods and services (Churski et al., 2021).

Finally the term *geographical* delimits its scope to spatially distributed phenomena. Given its multidisciplinary applications, multiple definitions of GIS coexist, each reflecting the priorities of different disciplinary traditions. For instance, according to Burrough (Isah et al., 2023), GIS can be understood as a collection of tools that assist professionals across diverse sectors in integrating geo–spatial and relational data. Mogorovich, by contrast, defines GIS as a system composed of three fundamental components (data, tools, and people) highlighting the active role of human expertise in transforming data into meaningful information (Mogorovich & Salvatori, 2022). Such definitions underscore not only the technical architecture of a GIS, but also its epistemological foundation as a sociotechnical system.

Far from being a neutral technology, GIS integrates computational capabilities with human judgment and disciplinary interpretation, shaping both the production and the understanding of geographic information. Representing and managing spatial information within GIS requires the construction of data modeling such (abstract) physical realities. These models should be sufficiently comprehensive to accommodate diverse real-world objects such as areas, lines, points, and elevations while remaining flexible enough to adapt to complex combinations of spatial entities (Zambon & Salvati, 2019).

Beyond geometric representation, a GIS should also manage information about the spatial relationship between entities—connectivity, adjacency, and containment (Howari & Ghrefat, 2021)—requiring explicit topological structuring of data to ensure that relationships are formally defined and computationally accessible. In addition, effective data models should integrate descriptive attributes (physical, ecological, and socioeconomic) that capture the multifaceted character of landscapes. Together, geometry, topology, and attributes, constitute the three essential components of data models, enabling GIS to function as a comprehensive system for spatial analysis, interpretation, and decision-making.

Historically, early GIS systems were primarily used for generating digital cartography, especially for geographic and geological applications (Elhashash et al., 2022), and for comparing thematic data distributions. Over time, statistical and computational integration became increasingly central, as standard packages began incorporating GIS technologies for Exploratory Data Analysis (EDA) and hypothesis testing (Ekbote et al., 2023). Yet, the statistical discipline had already developed a parallel tradition in geo-statistics, originating from Matheron's 'Theory of Regionalized Variables' (1971), which provided a framework analyzing phenomena continuously distributed across space (Boente et al., 2020). The ability to generate spatial models of continuous territorial variables via interpolation (or mathematical modeling) remains invaluable for natural sciences such as pedology, climatology, geography, and ecology.

In contexts characterized by high socioterritorial complexity, GIS technology assumes an equally critical role. It facilitates the development of advanced data processing and visualization techniques that offer three main advantages for applied research and spatial planning, namely (a) interdisciplinary integration, promoting collaboration among diverse socioterritorial disciplines, (b) a dynamic representation of social and environmental phenomena, linking them to ongoing territorial transformations, and (c) decision-making optimization, by integrating all relevant factors into comprehensive evaluations, thereby reducing overall complexity (e.g., Zambon et al., 2018).

Territorial systems are characterized by a dense network of interrelated variables, in which anthropogenic pressures generate cascading and often unpredictable effects. Effective governance of this complexity demands broad-spectrum territorial knowledge, achievable through the integrated analytical capabilities of GIS (Cárdenas et al., 2022). The conceptualization and implementation of a specialized GIS such as the one dedicated to landscape analysis raises several fundamental questions: "How should the system be designed?," "What data should be included, at what frequency, and over what spatial extent?" and "How should these data be structured, stored, and continuously updated?" The answers depend on the objectives and analysis' scale. For instance, global databases managed by international institutions generally operate at low resolution, covering vast regions with data on climate, population, and environmental conditions (Bansal, 2020). By contrast, socioenvironmental databases at regional or provincial scales provide high-resolution information about smaller territories, allowing for more detailed analysis (Zambon et al., 2017).

Two additional aspects should be emphasized in GIS construction (AlSaggaf & Jrade, 2023). First, the selection of information is dictated not only by project goals and administrative constraints but also by the availability and quality of data. Deficiencies often result from heterogeneous datasets and the lack of methodological comparability. Second, inconsistencies in data continuity over time often hinder the construction of robust historical series, which are essential for identifying trends, analyzing landscape evolution, and producing reliable predictive scenarios. In this regard, the development of specific socioenvironmental indicators represents a key operational priority in landscape analysis. Ecological indicators support various activities, such as estimating the magnitude of certain environmental processes, identifying causal factors, quantifying impacts, justifying corrective expenditures, and monitoring intervention effectiveness (Munafo et al., 2010; Ferretti et al., 2014; Fares et al., 2017).

Within a GIS framework, these analyses can be implemented through raster and vector data structures: the raster model organizes geospatial data as grids, whereas vector models represent discrete features points, lines, and polygons. Additionally, satellite imagery plays a pivotal role in monitoring landscape dynamics over time and across scales (Han et al., 2022). Ultimately, the effectiveness of any quantitative landscape analysis depends on selecting the most appropriate analytical tools, consistent with the objectives and the spatial–temporal resolution of available data (e.g., Ferretti et al., 2014). Equally important is the active involvement of local communities, whose knowledge of their territories enriches and validates analytical results. Community participation fosters greater environmental awareness, strengthens territorial identity, and enhances the sustainable management of rural and peri-urban landscapes. Thus, integrating quantitative methodologies with participatory approaches provides the most effective and holistic framework for contemporary landscape evaluation and management (Hasti et al., 2023).

A Functional Assessment of Landscapes

A fundamental framework in environmental studies and public policy for identifying effective strategies for sustainable resource management and mitigating global environmental impacts is the IPAT model (Zhang et al., 2024). This simplified theoretical construct explores the relationship between population dynamics, human well-being, and environmental impact (I). Specifically, it posits that environmental impact results from the combined effect of three key factors: Population (P), Affluence (A), and Technology (T). According to this model, the magnitude of environmental impact depends not only on the number of people inhabiting a given territory, but also on their level of consumption, and the technologies employed to sustain that consumption (Engström & Kolk, 2024). IPAT underscores that reducing environmental pressure requires coordinated interventions addressing all three factors simultaneously. Despite its conceptual clarity and simplicity, IPAT exhibits notable limitations. Key among these are the difficulty of empirically testing hypotheses derived from its strictly multiplicative structure and the assumption of proportionality among variables, which tends to oversimplify the complex (often non-linear) relationships characterizing real-world socioenvironmental systems. To overcome these limitations, the STIRPAT model (namely "Stochastic Impacts by Regression on Population, Affluence, and Technology") was introduced as a stochastic reformulation of IPAT equation (Krishnendu & Patra, 2025).

Stochastic Impacts by Regression on Population, Affluence, and Technology (STIRPAT) employs regression analysis to empirically investigate the effects of the driving forces (population growth, affluence, and technology) on environmental outcomes. This approach enables estimation of elasticities and testing of non-proportional relationships among variables, thus allowing for greater analytical flexibility. Through quantitative modeling, STIRPAT helps determining the relative contribution of each factor to overall environmental degradation, particularly with regard to greenhouse gas emissions, energy consumption, deforestation, and water resource depletion (Li et al., 2023). Subsequent refinements of the STIRPAT model have enhanced its analytical capacity and broadened its empirical applicability. These developments include the introduction of the concept of "ecological elasticity" (Lohwasser et al., 2020), which facilitates a more nuanced interpretation of model coefficients, and the integration of additional explanatory variables reflecting specific socioeconomic, institutional, or technological contexts. Through these advancements, the STIRPAT framework has evolved into a versatile tool for analyzing the determinants of environmental change and for empirically assessing the complex interplay among population, affluence, and technology.

The simplified architecture of STIRPAT can be effectively applied to a wide variety of environmental processes. One illustrative example is soil erosion, conceptualized within this framework as an environmental impact resulting from anthropogenic drivers such as population density and the technological means employed in land exploitation. Population acts as a fundamental driver influencing the intensity of land use and the economic activities that accelerate soil degradation. For instance, recurrent wildfires in densely forested regions lead to extensive losses of tree cover and biomass, exemplifying the compounded effects of human pressure and environmental feedbacks on ecosystem stability.

A number of quantitative models have been developed to assess soil erosion risk, among which the Universal Soil Loss Equation (USLE), remains one of the most widely applied frameworks (Thakuriah, 2023). The USLE estimates soil loss as a function of multiple variables, including slope gradient, vegetation cover, soil type, and rainfall intensity. Its improved (and implemented) version, namely the Revised Universal Soil Loss Equation (RUSLE), retains the same fundamental (mathematical) structure but introduces significant methodological improvements, extending its applicability to complex territorial and climatic contexts (Luvai et al., 2022). Similarly, the IPAT framework can be adapted to the study of desertification, integrating a broad set of interacting drivers. Population growth, for instance, intensifies pressure on finite natural resources such as water and soil, accelerating processes of land degradation (Nisa et al., 2021). Economic expansion and rising consumption levels often promote unsustainable agricultural practices including deforestation, overgrazing, and intensive cultivation that contribute to physical and chemical deterioration of soils. Inadequate or inefficient technological systems for resource management further exacerbate such effects, reinforcing the feedback mechanisms that drive desertification.

Within this context, the Environmentally Sensitive Areas (ESA) model (Niu et al., 2020) provides a fundamental framework for identifying, classifying, and managing areas that are particularly vulnerable to environmental degradation. The model's primary objective is to preserve natural habitats and protect endangered species, while simultaneously promoting the sustainable use of natural resources in regions at risk of degradation or emerging desertification (Ferrara et al., 2020). Areas delineated within ESA framework may be designated for specific conservation measures, regulatory restrictions, or incentive-based management programs aimed at ensuring their long-term ecological integrity (Table 6.1).

The ESA model is grounded in the evaluation of territorial zones where human activities exert significant environmental pressure, thereby necessitating the adoption of targeted protection and management strategies to mitigate adverse impacts. It was frequently implemented through national and international policy instruments and funding initiatives that promote sustainable resource management and biodiversity conservation. By guiding strategic planning and prioritizing the allocation of conservation resources, the ESA approach plays a crucial role in addressing global challenges such as deforestation, soil erosion, habitat fragmentation, and biodiversity loss. Ultimately, it contributes to the preservation and restoration of ecologically fragile landscapes, ensuring their resilience and continuity for future generations.

The considerable heterogeneity of environmental degradation processes, particularly those associated with the physical deterioration of land, is one of the most emblematic manifestations of ecological decline, and poses significant challenges for environmental monitoring and the development of effective mitigation strategies. Over the past two decades, however, renewed attention has been devoted to such issues, especially within the Mediterranean region, where the combined pressures of climate variability and anthropogenic activity have made land degradation a matter of urgent concern. Building on several European Community-funded research initiatives, various studies have empirically addressed land degradation as a complex, multidimensional phenomenon encompassing both the quantitative and qualitative (biophysical and/or economic) deterioration of one (or more) component(s) of natural capital over time. Increasingly, scholarly attention has shifted toward the systematic analysis of socioeconomic drivers that interact with natural capital, particularly soil and water resources, and that collectively influence degradation dynamics (Akhtar et al., 2021).

Table 6.1 An example of variables implemented in a generic Environmentally Sensitive Area (ESA) model, with details on measurement units and data sources for a Mediterranean country. (Source: Authors' elaboration of official statistics)

Size	Variable	Stairs	Measure	Source
Soil quality	Texture	1:250,000	Sensitivity class	Ministry of Agriculture
	Depth	1:250,000	Mm	Ministry of Agriculture
	Parental material	1:250,000	Sensitivity class	Ministry of Agriculture
	Skeleton	1:250,000	Sensitivity class	Ministry of Agriculture
	Drainage	1:250,000	Sensitivity class	Ministry of Agriculture
	Slope	1:25,000	%	Ministry of Environment
Climate quality	Average precipitation	1:500,000	Mm	Weather statistics
	Aridity index	1:500,000	mm/mm	Weather statistics
	Exposure	1:25,000	Angle	Ministry of Environment
Vegetation quality	Fire risk	1:100,000	Sensitivity class	Corine Land Cover
	Erosion protection	1:100,000	Sensitivity class	Corine Land Cover
	Drought resistance	1:100,000	Sensitivity class	Corine Land Cover
	Level of vegetation cover	1:100,000	Sensitivity class	Corine Land Cover
Land management quality	Population density	1:400,000	People km $^{-2}$	Population census
	Demographic variation	1:400,000	%	Population census
	Land use intensity	1:100,000	Sensitivity class	Corine Land Cover

These studies have progressively integrated theoretical frameworks with empirical methodologies, capturing the intricate feedback mechanisms linking human activity, environmental change, and territorial vulnerability. Despite considerable (scientific) progress, the relationship between sustainable development and land degradation remains only partially understood within the Mediterranean context (Halbac-Cotoara-Zamfir et al., 2020). The persistence of this knowledge gap highlights the urgent need for interdisciplinary research approaches that effectively integrate environmental, socioeconomic, and policy dimensions. Such integration is essential to disentangle the underlying causes of degradation, assess its long-term implications, and support the design of holistic mitigation strategies consistent with the objectives of sustainable land management and climate resilience.

Assessing Landscape Quality

The assessment of environmental quality within the landscape has acquired a central role in spatial planning, natural resource management, the designation of protected areas, and the broader field of environmental conservation. Interest in this topic spans multiple disciplines including ecology, nature conservation, forestry, economics, urban and regional planning, landscape architecture, and environmental sociology. Since the 1970s, scholars have proposed methodologies for evaluating quality and naturalness of sites of ecological interest, typically grounded in theoretical approaches, practical applications, and empirical data concerning soil use, land cover, and resource consumption.

The concept of environmental quality, particularly in relation to nature conservation and land-use planning, is now widely recognized and consolidated in recent literature. Two main approaches can be considered in environmental quality assessment, the one based on the functional value of ecosystems for human society (Environmental Value) and the other focused on the intrinsic properties of ecosystems themselves (Naturalness). Landscape quality is often assessed by identifying the direct benefits that ecosystems provide to humans (e.g., economic, cultural, social, historical, or aesthetic). This approach typically relies on subjective judgments, combining empirical observations with expert consensus based on predefined evaluation criteria. Environmental value is thus assigned using standardized rating scales, in which the cumulative scores of valued components yield an overall measure of environmental quality. A representative example of such value-based approaches can be found in forestry, where scholars quantify the aesthetic–functional value or scenic beauty of woods. These assessments exemplify the integration of perceptual and utilitarian dimensions in quantifying the human appreciation of natural systems.

Although widely used, these methodologies are typically grounded in standardized (yet subjective) evaluations of specific site parameters, as they assess environmental quality primarily from the standpoint of human perception, often excluding biophysical or ecological variables. Consequently, they estimate the degree of aesthetic appreciation associated with the (observable) physical components of forest ecosystems, interpreted in relation to visual appeal, recreational value, and tourism potential (Xu et al., 2024). Such approaches draw on techniques originally developed in North America, notably the Scenic Beauty Estimate (SBE) and the Law of Comparative Judgment (LCJ). According to the Aesthetic–Functional Quality (QEF) method, a representative sample of potential users is presented with visual stimuli (typically photographic images) depicting a variety of forests (Mundher et al., 2022). Participants are then asked to assign aesthetic–functional scores on a numerical scale, generally ranging from 1 to 10, for each image.

The dependent variable is computed as the mean appreciation score per image. Subsequently, a panel of experts identifies a set of independent variables describing forest-site characteristics that may influence aesthetic–functional judgments. Each variable, or descriptor, is assigned a score from 1 to 5, and these descriptors are subsequently aggregated into a smaller set of independent factors through averaging (or dimensionality-reduction) procedures. This process enables the construction of an interpretative model that correlates subjective appreciation values (QEF) with objectively measurable forest attributes, thereby linking perception and structure.

Beyond perception-based methods, a second group of approaches focuses on biophysical variables derived primarily from land-use mapping and analysis of land-use change over time. Within these frameworks, standardized naturalness scales are commonly applied to classify and compare environmental conditions. Three main types of measurement scales are typically used: (a) simplified naturalness scales (assigning schematic "degrees of naturalness" based on overall assessments of vegetation, flora, and the degree of human modification), (b) multivariate scales (constructed from multiple parameters (e.g., rarity, diversity, climatic conditions, structural complexity, and anthropization), each independently evaluated and summarized into a composite index), and (c) landscape-based scales—assigning a naturalness value to each mapping unit or plot, followed by calculation of percent coverage to obtain an aggregate evaluation of the entire area. These scales provide structured, replicable frameworks quantifying landscape naturalness and environmental quality, thus facilitating cross-site comparisons and informing conservation and land-management decisions.

A representative example of this multiscalar approach is the Priority Conservation Index (IPC). Their method identifies areas of high natural value within a given territory, supporting the prioritization of conservation interventions (Wani et al., 2022). Although the IPC can incorporate biological diversity rather than land cover alone, the authors underscore the central role of vegetation in determining environmental quality. To ensure that the resulting index provides actionable information for conservation planning, each parameter was weighted according to its relative ecological importance. The IPC integrates variables related to environmental characteristics, site naturalness, biogeographical significance, biological diversity, and anthropogenic pressure. Attributes associated with rarity and human disturbance are assigned higher weights than those reflecting general environmental variety.

The weighted scores are summed to produce a final conservation priority value for each area. The level of environmental quality reflects an ecosystem's intrinsic capacity for self-preservation. This notion introduces the related concepts of "naturality" and "artificiality." Within this framework, land-cover types were investigated in terms of their constituting elements, and models were developed to correlate indicators such as biotic diversity, rarity, and species richness with overall environmental quality. Comparing maps of current and potential vegetation can reveal the degree of artificiality of different vegetation types—the greater the deviation from potential (climatically stable) vegetation, the higher the level of human alteration.

Similarly, it was introduced a quantitative methodology assessing the environmental quality of forests, emphasizing objective parameters over perceptual evaluations. The method was tested on four experimental plots exhibiting various degrees of human disturbance from near-natural forests approximating potential vegetation to highly degraded sites. Through this approach, scholars identified and prioritized key environmental quality indicators, assigning relative weights without resorting to subjective interpretation. Over the past three decades, a broad spectrum of methodologies has emerged for the measurement of environmental values, differing in scope, precision, and purpose. When comparing the two paradigms concerning the evaluation of ecosystemic naturalness and the estimation of environmental value (e.g., Sannigrahi et al., 2020) it becomes evident that methods based on ecosystemic naturalness are more objective, as they rely on biophysical parameters rather than perceptual assessments.

However, these approaches are often constrained by time, cost, and methodological complexity required to generate detailed vegetation maps. In contrast, value-based techniques, though faster and more accessible, are inevitably more subjective. Ultimately, the choice between these approaches depends on the objective of the study, available resources, and the desired level of analytical precision. For comprehensive environmental evaluations, the integration of both methodologies combining the objectivity of naturalness assessments with the flexibility of perceptual valuation offers a balanced and effective framework for landscape analysis, environmental monitoring, and conservation planning.

Qualitative Landscape Analysis

Analyzing a landscape from a qualitative perspective—namely assessing the quality and condition of all its constituent components—represents a crucial phase in the environmental decision-making process. Understanding the aesthetic, ecological, and cultural dimensions of landscape quality not only supports policies aimed at preserving environmental integrity and protecting natural and cultural areas from potential threats, but also provides a valuable foundation for land-use planning (Mounsey et al., 2022). Qualitative landscape analysis constitutes a methodological approach that enables the identification (and evaluation) of areas of high environmental and cultural value, as well as those that are fragile or vulnerable, and thus require careful management to ensure sustainable development.

By distinguishing resilient from sensitive zones, such analysis supports balanced spatial planning and the formulation of targeted conservation strategies. At the same time, qualitative landscape assessment contributes significantly to the valorization of cultural heritage. When implemented comprehensively, it promotes the recognition, documentation, and enhancement of sites of historical and cultural importance.

In this way, qualitative analysis functions simultaneously as an environmental and cultural instrument, integrating aesthetic, ecological, and heritage dimensions to advance sustainable territorial governance and the long-term protection of landscape identity. Analyzing landscape quality also makes it possible to assess the potential impact of human activities on the surrounding environment, thereby supporting the adoption of mitigation measures to reduce adverse effects.

In this sense, qualitative landscape analysis constitutes an essential tool for sustainable and informed management of both natural and cultural resources. A variety of methodological techniques can be applied to run qualitative landscape analysis. Among the most widely used is the survey method, based on the release of open- and/or closed-ended questionnaires designed to collect responses from participants through individual interviews. To conduct a survey effectively, operators should define objectives, followed by the design of a questionnaire to be administered to a representative sample of respondents. The questionnaire comprising a structured set of questions facilitates the systematic collection of relevant information, supporting hypothesis testing and qualitative–quantitative data analysis. Specifically, this tool enables the collection of information regarding knowledge, behaviors, attitudes, opinions, motivations, and perceptions.

One of the main advantages of the questionnaire lies in its flexibility: its structure can be adapted to diverse research objectives. Each question type (whether open, closed, structured, multiple-choice, hierarchical, or retrospective), is suited to capturing specific categories of information. Moreover, the questionnaire is cost-effective and can be administered in multiple formats (paper-based, electronic, postal, or telephonic) allowing full data comparability. Despite these advantages, survey-based methods also have notable limitations. The absence of an interviewer, for instance, can lead to misinterpretation of topical questions, as participants lack the opportunity for clarification. Operators should therefore design questions that are clear, unambiguous, and consistent with the objectives of the study. However, the pursuit of interpretive uniformity may constrain the range of possible responses.

Another potential source of bias is the "emphasis effect," which occurs when particular elements of the questionnaire are inadvertently highlighted, influencing subsequent answers. Once the questionnaire is finalized, an essential step is the identification of a representative sample. Unlike a census, which encompasses the entire population, a sample includes only a fraction, introducing a sampling error that reflects a degree of uncertainty in extrapolating results. Sampling error tends to increase as sample size decreases, though it can be minimized through appropriate sampling design. Ideally, sampling should be probabilistic, ensuring that respondents are randomly selected from a defined population list. In some cases, a non-probabilistic sampling could be used, guided by prior knowledge or demographic data, though this approach limits generalization.

Another widely adopted research technique is the individual interview, which typically employs open-ended and non-hierarchical questions. Interviews are conceived as dialogic exchanges between researchers and respondents capable of providing relevant information about the topic under study. This method allows for the collection of in-depth, context-rich data, as interviewers can adapt their questions dynamically in response to participants' answers. The interactive nature of interviews fosters rich communication, improves data quality, and allows for the clarification of ambiguities. However, interviews are also subject to interviewer bias, where tone, phrasing, or emphasis may inadvertently influence responses. Furthermore, interviews can be time-consuming and resource-intensive, especially when involving large samples, and the potential for incomplete or inaccurate responses may affect reliability. Interview-based research typically proceeds through several stages: the design of an interview guide, selection of respondents, recording and transcription of conversations, and qualitative data analysis.

When rigorously implemented, this method yields valuable insights into perceptions, motivations, and social dynamics that are difficult to capture through quantitative approaches. To ensure reliable results, the researcher should minimize question complexity and maintain clarity in language and delivery. Once participants are selected according to predefined criteria, logistical aspects such as timing, location, and mode of communication (in-person, telephone, or video conference) should be established. The interview should be conducted professionally and respectfully, with careful attention to accurate recording or transcription of responses.

Collected data should be analyzed objectively, with conclusions drawn from empirical evidence. Nevertheless, as with any empirical method, measurement and interpretation errors may arise from respondent bias, question ambiguity, or recording inaccuracies. Beyond empirical observation, it is essential to examine the theoretical foundations that underpin landscape phenomena. A thorough bibliographic and conceptual review allows operators to understand the interactions among environmental, cultural, and socioeconomic variables. In the context of landscape research, theoretical analysis focuses on the processes of landscape formation, organization, and transformation arising from the interplay between natural dynamics and human activity (Shores et al., 2024).

Such theoretical inquiry often integrates frameworks from related disciplines (land-use planning, environmental geography, ecology, and landscape architecture) enabling an interdisciplinary understanding of how cultural, historical, and economic forces shape landscapes. Through this synthesis, scholars can develop conceptual and analytical models to predict and manage landscape change, assess human impacts, and inform sustainable development policies. Furthermore, comprehensive landscape analysis should integrate quantitative and qualitative approaches using objective (comparative) criteria and involving multiple representative case studies.

Comparative analysis serves to identify patterns of similarity and divergence among landscapes, improving understanding of natural dynamics, driving forces, and ecological outcomes. This requires selecting cases that share key characteristics such as vegetation type, climate, geology, and land use, applying analytical tools to interpret their relationships. The principal aim of comparative landscape analysis is to identify the determinants of landscape structure and functions, thereby guiding the design of management, protection, and restoration strategies. These methodologies (surveys, interviews, theoretical inquiry, and comparative analysis) represent some of the main approaches for conducting qualitative landscape analysis (Lester et al., 2020).

As with quantitative methods, it is vital to select the approach best suited to the research context, data availability, and analytical objectives. Survey-based techniques, for instance, provide valuable insights into the perceived quality of landscapes, yet remain constrained by subjectivity, data quality, and the complexity of landscape systems. Moreover, such methods may have limited capacity to capture the temporal variability and spatial heterogeneity of dynamic landscapes, particularly those exhibiting high biological and ecological diversity. Ultimately, integrating qualitative and quantitative approaches combining subjective perception with objective measurement offers the most comprehensive pathway for understanding and managing landscape complexity, ensuring that environmental, social, and cultural dimensions are jointly considered in sustainable territorial governance.

The Economic Evaluation of Landscapes: A Primer

Market generally provides individuals with clear and quantifiable information upon which to base their decisions, as most goods are tangible, their characteristics are well-defined, and their prices are explicitly determined through market transactions. However, this conventional mechanism does not extend to environmental goods and services, which typically lack a direct market price and are therefore subject to valuation uncertainty, particularly with respect to their intrinsic, non-market, or non-use values, such as those associated with landscape quality, biodiversity, or cultural heritage. Moreover, many environmental assets exhibit the characteristics of public goods: they are non-excludable, meaning that individuals cannot be prevented from benefiting from them, and non-rivalrous, meaning that one person's enjoyment does not diminish another's. These features generate market failures, as private markets are unable to efficiently allocate such goods or reflect their true social and/or ecological value.

Consequently, individuals and institutions lack appropriate market signals to guide environmental decision-making, often resulting in overexploitation, conservation underinvestment, and unsustainable resource use. To address this deficiency and support evidence-based environmental policy, it becomes necessary to assign an economic value to environmental goods and services. This enables the comparison of alternative policy options and management strategies within a coherent cost–benefit framework. In response, environmental economics has developed a range of valuation techniques that estimate these values in the absence of explicit market prices.

Most of these methods are grounded in the concept of Willingness To Pay (WTP), the maximum amount an individual is willing to pay to obtain a specific environmental improvement or to avoid its deterioration (Pouta et al., 2002). This framework allows for monetary estimation of environmental benefits even in contexts where market information is limited or nonexistent, thereby providing critical inputs for natural resource management, policy evaluation, and sustainability assessment. The accuracy of such valuations depends on contextual factors, including socioeconomic conditions, cultural norms, and individual preferences and perceptions. A central conceptual foundation in this field is Total Economic Value (TEV) framework, which captures the full spectrum of benefits derived from environmental goods.

Total Economic Value (TEV) encompasses both use values including direct and indirect uses and non-use values, which integrate option, existence, and bequest values. This classification provides a comprehensive understanding of the multiple ways in which the environment contributes to human welfare, beyond the boundaries of market exchange. Closely related to these concepts is consumer surplus, defined as the difference between the maximum amount a consumer is willing to pay for a good and the market price actually paid. In environmental economics, consumer surplus serves as an indicator of the aggregate perceived value of environmental improvements or conservation interventions.

However, it may not fully capture broader social costs or externalities associated with degradation, such as pollution, habitat loss, or ecosystem decline. Consequently, monetary measures alone may underestimate the true societal importance of environmental assets (Abdullah et al., 2011). The economic valuation of environmental resources thus represents an innovative and evolving research field, marked by ongoing theoretical and methodological challenges. Reliable assessments require explicit recognition of the multidisciplinary nature of environmental analysis, integrating insights from ecology, economics, geography, and the social sciences.

Moreover, the estimation of environmental damages and benefits demands the development of robust, transparent, and context-sensitive methodologies. Only through such integrated and interdisciplinary approaches can accurate, policy-relevant, and ethically grounded evaluations of environmental value be achieved. Although a universally accepted terminology has yet to be established, the theoretical underpinnings of environmental valuation are rooted in classical economic theory concerning the origin of value. According to this framework, individuals attribute worth to goods and services based on the subjective utility (or satisfaction) derived from their use or consumption. This conceptual foundation informs contemporary valuation methodologies, and provides the analytical basis for estimating the aggregate economic value of environmental assets (Isacs et al., 2021).

Within TEV framework, a fundamental distinction is drawn between use values and non-use values, which together represent the comprehensive economic worth of environmental assets. Use values can be classified as either direct or indirect. Direct use values refer to tangible and measurable benefits, such as revenues from timber production, agricultural yields, or recreational activities. Indirect use values encompass ecosystem services that contribute to human welfare in less visible ways, including carbon sequestration, climate regulation, and water purification provided by ecosystems such as forests and wetlands. Within this category, option values represent individuals' WTP to preserve the potential for future use of environmental resources, while bequest or legacy values capture the WTP to conserve environmental quality for the benefit of future generations. Non-use values, in contrast, are not associated with any actual or potential use of a given environmental resource. They reflect the intrinsic or existence value attributed to ecosystems, species, or landscapes, independent of human utility. These values express the ethical, cultural, and moral dimensions of human–nature relationships, and are grounded in the recognition of the inherent worth of nature, regardless of its economic utility.

The Total Economic Value of an environmental asset is generally conceptualized as the aggregate of use, option, and existence values. In practice, however, this aggregation often remains partial, as certain ecosystem functions such as biodiversity maintenance, nutrient cycling, and aesthetic appreciation, are inherently difficult to monetize (Plottu & Plottu, 2007). To address this limitation, several scholars have proposed substituting the term *non-use values* with passive use values (Carson et al., 1999), thereby emphasizing that even apparently "non-utilitarian" values often involve implicit forms of use, preference, or benefit. Within the field of environmental economics, a key methodological distinction is drawn between demand curve approaches and non-demand curve approaches, both essential for the monetary assessment of environmental goods and services (Eshet et al., 2006). These methods are grounded in microeconomic consumer theory, deriving environmental values from individuals' expressed or revealed preferences:

- Expressed preference methods, including the Contingent Valuation Method (CVM) and Choice Modeling (CM), rely on hypothetical market scenarios in which individuals explicitly state their WTP or Willingness to Accept (WTA) compensation for specific environmental changes.
- Revealed preference methods, such as the Hedonic Pricing Method (HPM) and the Travel Cost Method (TCM), infer values indirectly by analyzing observed market behaviors, identifying the implicit prices individuals assign to environmental attributes through actual purchasing or travel decisions.

These methods are not based on preference data or utility functions; rather, they rely on biophysical or cost-based relationships to estimate environmental values:

- The Dose–Response method quantifies environmental damage by linking pollutant exposure to measurable biological or ecological effects.
- The Replacement or Restoration Cost Method estimates value based on the costs of restoring or substituting lost environmental services.
- The Averting or Defensive Behavior Method measures expenditures undertaken to prevent or mitigate environmental harm.
- The Opportunity Cost Method evaluates the benefits forgone when a resource is devoted to one environmental use rather than an alternative.

Although less directly connected to welfare economics, non-demand curve approaches are particularly useful in data-scarce contexts, or when the goal is to assess physical damages, restoration costs, or policy trade-offs. When combined with preference-based methods, these approaches yield a more comprehensive and policy-relevant understanding of the economic dimensions of environmental change. Ultimately, the economic valuation of environmental goods and services bridges the gap between ecological processes and economic decision-making. By quantifying the benefits of conservation and the costs of degradation, valuation methodologies provide a framework for integrating environmental values into public policy, land-use planning, and sustainable development strategies. Yet, it remains essential to recognize that not all environmental values can (or should be) expressed in monetary terms. A balanced and ethical approach that combines economic

Table 6.2 Summary of evaluation methodologies outlining advantages and disadvantages in their empirical application (see the main text for abbreviations' use). (Source: Authors' elaboration of recent literature)

Value category		Indicator	Method	Advantages	Disadvantages
Use	Cognitive (scientific, historical)	Archaeological heritage, historical museums	Market-based methods	Using market prices	Not applicable in the absence of prices and rates; only takes into account use values
	Recreational (social)	Transport cost, entrance ticket	CVM	Relative simplicity of data collection	It does not take into account the benefits of travel and conservation of the asset; it does not isolate the cultural components of the benefit of travel
	Aesthetic	Transport cost, entrance ticket	CVM, Hedonic pricing	Capturing the non-market attributes of goods	There is no perfect real estate market, high survey costs, possible distortions in responses, difficulties in data analysis
No use	Existence, option, legacy	Conservation of cultural heritage	CVM	Ability to capture non-market attributes	Difficult to quantify from a monetary perspective; influence of social and personal values

valuation, ecological integrity, and cultural recognition is crucial for advancing a truly sustainable and equitable relationship between human societies and natural environments (Table 6.2).

Dose–Response Approaches

The Dose–Response Method (Poorter et al., 2010) represents a fundamental analytical approach for evaluating how the biological responses of plants, animals, or humans vary as a function of the concentration (or dose) of a specific pollutant in the environment under investigation. The technique involves the controlled exposure of organisms to a range of pollutant concentrations, followed by systematic measurement and statistical analysis of the corresponding biological or physiological effects. The primary objective of this method is to establish a quantitative dose–response relationship, also referred to as a dose–response function, which expresses the relationship between the magnitude of exposure to a contaminant and the resulting biological or ecological response (Moffett et al., 2022).

These responses may include adverse human health effects, reductions in plant growth, alterations in animal behavior, or more complex forms of ecosystem imbalance. By identifying threshold levels, that is, exposure limits beyond which measurable harm occurs, the dose–response method provides essential data for risk assessment in both environmental and public health contexts. Insights derived from this analysis are foundational for defining environmental quality standards and safety thresholds for air, water, and soil. The results are subsequently incorporated into regulatory frameworks and policy instruments aimed at preventing or mitigating the negative impacts of pollution on ecosystems and human well-being (Cox Jr, 2020).

Implementing a dose–response study requires the collection of high-quality empirical data on pollutant concentrations and their corresponding biological (or ecological) effects, supported by rigorous sampling designs, laboratory protocols, and analytical procedures. Once collected, these data are used to determine the mathematical form of the dose–response function and to estimate the critical concentration or exposure level at which statistically significant adverse effects occur. Establishing this relationship provides a scientifically robust foundation for quantifying environmental damage and developing evidence-based risk management strategies. A practical illustration can be drawn from studies on air pollution and agricultural productivity (Sillmann et al., 2021). Suppose empirical analyses reveal a statistically significant relationship between rising concentrations of airborne pollutants such as sulfur dioxide (SO_2) or ozone (O_3), and corresponding reductions in crop yields. In this case, the dose–response method enables scholars to estimate both the extent of yield loss attributable to pollution and the economic implications stemming from changes in market supply, agricultural income, and commodity prices (Osborne et al., 2016).

By integrating biophysical data (measured crop yield reductions) with economic indicators (market valuation of agricultural output), it becomes possible to assess the monetary value of environmental degradation. Such integrated analysis provides a robust empirical basis for designing policy interventions targeting pollution abatement, sustainable resource management, and the promotion of environmentally responsible agricultural practices. When the analysis is extended to human health, the dose–response framework becomes considerably more complex and multidimensional. In this context, it is necessary to account for variables such as individual health status, exposure duration and intensity, population vulnerability, and differential susceptibility to pollutants, all of which influence the overall magnitude of pollution-related impacts (Lesser et al., 1990).

The assessment of these effects requires consideration not only of the risks of morbidity and premature mortality attributable to pollutant exposure, but also of the economic valuation of human life and health within a welfare-economic framework. To quantify such impacts, environmental economists and public health scholars employ indicators such as the Value of a Statistical Life (VSL) and the Value of a Statistical Life Year (VSLY). These measures estimate society's Willingness to Pay (WTP) to reduce the probability of mortality (or morbidity) associated with environmental hazards. Although ethically sensitive and methodologically contested, these metrics provide essential quantitative inputs for cost–benefit analysis, and for evaluating the economic consequences of regulatory inaction. They are widely applied in the formulation of pollution control strategies, health protection standards, and environmental mitigation policies at both national and international levels, offering a systematic means to integrate human health outcomes into broader assessments of environmental and economic sustainability.

In essence, the dose–response approach represents the analytical foundation for establishing the relationship between variations in pollutant concentrations and the resulting physiological and epidemiological responses. When integrated with economic valuation techniques, it enables policymakers to assess the broader social and economic consequences of environmental degradation, encompassing its effects on human well-being, agricultural productivity, and ecosystem integrity. Ultimately, this interdisciplinary framework provides a rigorous basis for evidence-based environmental governance, supporting the formulation of policies and regulatory instruments designed to mitigate the adverse impacts of pollution on both human health and the natural environment.

Replacement Cost and Opportunity Cost

The Replacement Cost Method, also referred to as the Restoration Cost Technique, estimates the value of an environmental good (or service) by calculating the costs required to restore or replace a damaged natural asset (Wyatt, 2009). These expenditures are interpreted as proxies for the economic benefits derived from restoration, under the assumption that society is willing to pay at least an equivalent amount to recover the lost ecological functions and services. In essence, the method posits that the restoration cost represents the minimum monetary value of the environmental damage. However, this approach requires considerable methodological caution, as it is theoretically valid only under specific regulatory or planning conditions that render restoration activities obligatory rather than discretionary.

For instance, when a legally binding environmental quality standard such as a minimum allowable threshold for water quality (e.g., Jackson et al., 2014) is in force, the costs necessary to achieve compliance can be regarded as an approximate monetary estimate of the environmental benefits associated with meeting that standard. Similarly, in contexts governed by broader sustainability constraints that prohibit further degradation of environmental resources, the replacement cost may serve as a lower-bound estimate of the value of the incurred damage. A clear example can be observed in wetland ecosystems, many of which were historically degraded but are now protected under international frameworks such as the Ramsar Convention on Wetlands (Matthews, 1993).

Financial resources required for their restoration, whether complete or partial, can be interpreted as measures of the economic benefits of conservation, or conversely, as the costs of environmental loss. Within this context, the so-called *shadow project approach* operates on the assumption that restoration costs, given sustainability constraints, represent a minimum estimate of environmental damage. In more practical terms, consider the case of urban noise pollution: residents may choose to install sound-insulating materials in their homes as a mitigation strategy. The expenditures associated with such interventions can be used as indicators of the value attributed to a safe environment, particularly when direct mitigation at the source is not feasible. Thus, the Replacement Cost Method provides a useful tool for environmental and economic assessment, enabling policymakers to make informed decisions regarding the sustainable management of natural resources and the efficient allocation of restoration funds (Halls, 1997).

Nevertheless, this method presents certain limitations. It should not be applied without considering the broader social, ecological, and cultural benefits associated with restoration, which often exceed the mere financial costs of intervention. Complementarily, the Opportunity Cost Approach evaluates the benefits forgone when environmental degradation occurs as a result of alternative land uses. For instance, the drainage of wetlands to promote intensive agriculture may yield short-term economic gains but simultaneously entails the loss of vital ecosystem services such as flood regulation, biodiversity maintenance, and carbon sequestration. In such cases, comparing the foregone environmental benefits with the economic returns of alternative development provides insight into whether the trade-off is socially and environmentally sustainable.

Unlike methods that directly estimate the benefits of environmental goods and services, the Opportunity Cost Approach focuses on the economic costs of activities that degrade or deplete natural resources (Borrego & Skutsch, 2014). Rather than measuring the value of environmental improvements, it assesses the trade-offs associated with resource allocation to environmentally harmful uses, thereby identifying the potential welfare gains achievable through more sustainable alternatives. This approach has proven particularly valuable in environmental policymaking (Westhoek et al., 2012), as illustrated in the case of wetland loss in the old continent, possibly associated with the Common Agricultural Policy (CAP).

Historically, the generous subsidies allocated for land reclamation and intensive agricultural expansion encouraged the drainage and conversion of wetlands, resulting in significant ecological degradation and a socially inefficient outcome. The subsequent reduction (or removal) of these subsidies effectively curtailed such environmentally damaging activities, demonstrating the policy relevance of opportunity cost analysis in guiding reform.

The Opportunity Cost Approach thus represents a powerful instrument for environmental decision-making and spatial planning, as it enables policymakers to evaluate the implicit environmental costs of competing land uses (Blanco et al., 2016) and to design corrective measures that promote more sustainable resource management. Empirical evidence from case studies confirms its effectiveness: the application of opportunity cost principles has informed the development of planning tools and regulatory adjustments that have yielded tangible improvements in environmental conservation and ecosystem restoration outcomes.

Contingent Valuation

The Expressed Preferences Method, also known as the Stated Preferences Approach (Kroes & Sheldon, 1988), is based on the direct elicitation of individuals' preferences through structured questionnaires administered to a representative sample of respondents. Within this class of valuation techniques, the Contingent Valuation Method (CVM) represents the most widely applied and extensively validated approach. Similar to the Hedonic Pricing Method—which will be discussed later—CVM is grounded in the principle that individuals are capable of expressing their valuation of environmental goods and services. However, unlike hedonic approaches, which infer values indirectly from observed market behavior, contingent valuation relies on hypothetical market scenarios specifically designed to elicit individuals' Willingness to Pay (WTP) or Willingness to Accept (WTA) compensation for changes in the quality (or availability) of environmental goods.

In practical applications (e.g., Walker & Mondello, 2007), respondents are presented with a detailed description of hypothetical scenarios involving environmental improvement or degradation (e.g., wetland conservation or air pollution abatement) and are asked to state the maximum amount they would be willing to pay to secure the benefit or avoid the loss. This process allows researchers to derive a monetary measure of the perceived value of the environmental asset under consideration. The aggregate social value is then estimated by calculating the average WTP among respondents and extrapolating this figure to the entire population that benefits from the environmental good or service at stake.

By directly capturing individual preferences related to environmental quality, the Contingent Valuation Method provides a flexible and comprehensive framework for estimating both use and non-use values, including aesthetic, cultural, and existence values that are typically excluded from traditional market-based assessments. One of the principal advantages of CVM is the capacity to assign an economic value to environmental goods and services that are not traded in markets and may be physically or geographically inaccessible (Zhang et al., 2003). This is particularly relevant in cases such as the protection of Antarctica, where individuals often express a relatively high WTP for conservation despite having little or no intention of visiting the area directly. Such results illustrate the method's ability to capture non-use values, including existence and bequest values, which remain outside the scope of market-based valuation techniques.

Despite its broad applicability, CVM presents several well-known limitations. A primary concern involves the systematic underestimation of WTP. Since the method relies on individuals' declared monetary valuations, it assumes that survey responses accurately reflect the true value respondents assign to environmental resources. Empirical research and behavioral experiments have shown, however, that respondents frequently report lower WTP values than they would actually pay in real-world contexts—an effect attributed to hypothetical bias and strategic behavior.

Another well-documented issue is the discrepancy between WTP and WTA. The valuation question may be framed in two alternative ways:

How much are you willing to pay to obtain this environmental improvement? or

How much compensation would you require to forgo this environmental good or accept its degradation?

In theory, both measures should yield equivalent results under standard economic assumptions. In practice, however, WTA estimates are consistently higher than WTP values, reflecting loss aversion and the endowment effect, whereby individuals place greater value on goods they already possess or believe they are entitled to. Consequently, CVM captures stated preferences rather than revealed behaviors, introducing uncertainty into the accuracy and policy relevance of empirical results (Baymuminova et al., 2023). Recent studies have confirmed that individuals tend to perceive the cost of a loss as greater than the benefit of a gain, suggesting that the divergence between WTA and WTP may, paradoxically, reinforce the psychological validity of contingent valuation outcomes. However, such asymmetries can increase responses' variability, potentially resulting in significant underestimation of environmental values when relying solely on WTP.

Another methodological challenge concerns the scope of valuation that is, whether respondents are asked to evaluate an entire environmental asset or a portion of it. Empirical evidence suggests that individuals often declare similar WTP values when asked to value a subset of environmental goods (e.g., one lake within a system) and the entire resource (the full set of lakes). This phenomenon likely reflects budget allocation effects and anchoring behaviors.

To mitigate this bias, researchers may adopt a two-step approach, first asking respondents to estimate their total environmental budget and then to allocate portions of it to specific components. Alternatively, CVM can be applied to composite environmental assets while explicitly reminding respondents of budget constraints. A further concern relates to the payment vehicle, or the mechanism through which contributions are hypothetically made. The credibility and realism of this payment mechanism (e.g., taxes, entrance fees, or utility bills) strongly influence respondents' stated WTP. Providing realistic, secure, and context-appropriate payment options is therefore essential to eliciting reliable valuations.

Finally, CVM results may be affected by the starting point bias, which occurs when respondents adjust their stated WTP according to an initial reference value or suggested amount. Although modern survey design has largely abandoned the use of starting bids, this issue highlights the broader challenge of ensuring that elicited values are not artificially influenced by framing effects or question design. In summary, while the Contingent Valuation Method offers a powerful and adaptable tool for quantifying environmental values particularly for non-market goods, it requires careful design, testing, and interpretation to avoid systematic bias. When properly implemented, CVM provides essential (empirical) evidence for environmental policy, cost–benefit analysis, and sustainable resource management.

Travel Cost Approach

The Revealed Preference Method relies on the analysis of individuals' actual market behavior to infer the value they assign to environmental goods and services. In this framework, the Travel Cost Method (TCM) is among the most widely applied techniques for estimating the economic value of recreational sites (Das, 2013). The underlying principle of the method is straightforward: the costs incurred to visit a site are treated as a proxy for the recreational value that individuals derive from it. To implement the Travel Cost Method, surveys or questionnaires are used to collect information on visitors' points of origin and travel behavior. By analyzing empirical data, it becomes possible to estimate individual travel costs and relate them to the frequency of visits over a given period, typically one year.

Since visitation patterns depend on factors such as distance, accessibility, income, and sociodemographic characteristics, these variables are incorporated into econometric models to estimate the demand function for the recreational site. This demand curve describes the relationship between the cost of visiting and the number of visits, which constitutes the basis for quantifying the site's total recreational value. Once the demand curve is established, the annual recreational value of the site can be approximated by multiplying the average cost per visit by the total number of visits. The logic of the method thus mirrors that of traditional consumer theory, where the area under the demand curve represents the aggregate consumer surplus that is, the total net benefit individuals derive from accessing the site. Although the TCM appears relatively simple and intuitive, its application involves several methodological challenges.

One of the most significant concerns relates to the treatment of time as either a cost or a benefit. While the method assumes that travel expenditures reflect the value of recreational experiences, time spent traveling also carries an opportunity cost, as it could be devoted to alternative (productive or leisure) activities. Failure to include this component leads to a systematic underestimation of the site's total recreational value (Zhang et al., 2015). However, for some individuals, travel itself may be perceived as an enjoyable experience rather than a cost. For instance, scenic drives or hikes to reach a natural area may constitute part of the recreational experience.

In such cases, the positive utility derived from travel should be subtracted from the total cost, since it represents an additional benefit, not an expense. Another important limitation concerns multidestination trips. Visitors may include several recreational sites within the same journey, making it difficult to allocate travel expenses precisely among them. Although the total cost of the trip can be measured, only a fraction of that expenditure corresponds to the specific site under analysis, introducing uncertainty into value estimates. A related issue arises from the existence of substitute sites. Some individuals may visit a particular location because they genuinely value its specific attributes, while others may do so simply because no better alternatives are available.

The standard TCM framework may fail to capture this heterogeneity in visitor motivations, leading to overestimation or underestimation of the true recreational value of a given site (Clough & Meister, 1991). Furthermore, in some cases, individuals who highly value a recreational area may decide to purchase a residence nearby. As a result, their travel costs diminish substantially, and the TCM would consequently underestimate their actual willingness to pay for proximity to the site. Similarly, the presence of non-paying visitors, such as local residents who access the area on foot and incur no travel expenses, poses an additional challenge.

Despite facing negligible travel costs, these individuals may attribute very high recreational or existence value to the site values that the TCM fails to capture. While being largely sensitive to such limitations, the Travel Cost Method remains a powerful revealed preference technique for assessing the use value of natural and recreational resources. By analyzing real-world behavior rather than hypothetical scenarios, it provides empirical evidence of individuals' preferences for environmental goods. However, for accurate application, it requires careful treatment of time valuation, trip purpose, substitute sites, and local users, as well as robust survey design and econometric modeling.

Hedonic Valuations

The Hedonic Pricing Method (HPM) assigns economic value to environmental services that directly influence market prices, particularly those reflected in the real estate sector. In practice, housing prices are determined by a combination of structural, locational, and environmental attributes such as the number of rooms, proximity to workplaces, size of outdoor space, and the overall environmental quality of surrounding areas. The fundamental assumption of the method is that market prices incorporate individuals' implicit valuations of these attributes, allowing the environmental component of property value to be isolated and quantified. To apply HPM, scholars should collect detailed data on the multiple variables influencing property prices, including proximity to amenities, exposure to noise or pollution, landscape aesthetics, and the presence of nearby industrial facilities (Ruankham, 2025). Owing to these features, the method has traditionally been used to assess environmental costs (such as those associated with pollution, noise, or visual degradation) rather than to evaluate environmental benefits.

Historically, data collection for hedonic analysis was a labor-intensive and time-consuming process, often reliant on manual surveys. However, recent technological advances have considerably enhanced the accuracy and efficiency of this method. In particular, GISs have transformed data acquisition and spatial analysis, enabling precise mapping of environmental attributes and their proximity to residential areas (Lü et al., 2019). GIS tools facilitate the calculation of distances, the identification of environmental gradients, and the integration of topographic or land-use information, all of which contribute to more accurate estimations of how environmental factors affect property values (Doğan & Eroğlu, 2021). Consequently, GIS-based applications have improved the robustness of hedonic analysis, allowing for a more precise attribution of monetary value to landscape and environmental characteristics (Noor et al., 2015).

Despite its analytical power, the Hedonic Pricing Method presents several methodological and practical challenges. First, estimating the relationship between property prices and environmental quality requires advanced statistical modeling skills and detailed datasets. The complexity of housing markets combined with the interdependence of influencing variables introduces a high risk of model misspecification and potential bias in the estimation of implicit prices. Second, the HPM rests on the assumption that buyers can freely choose among different combinations of property attributes in accordance with their preferences. In reality, this choice is constrained by income levels, budget limitations, and the availability of suitable properties. Furthermore, external macroeconomic factors such as tax incentives, government regulations, mortgage interest rates, and broader market cycles extensively shape real estate markets (Lake et al., 2000). These exogenous influences complicate the analysis and interpretation of results, as they may obscure or amplify the environmental component of property values.

In summary, the Hedonic Pricing Method remains one of the most rigorous (revealed preference) approaches for valuing environmental quality through observable market transactions (Hidano, 2002). However, its successful implementation requires extensive, high-quality data and sophisticated econometric techniques. Like all environmental valuation methods, it presents both strengths and limitations. Therefore, the choice of valuation technique should be guided by the specific objective of the study, the availability of reliable data, the technical expertise of the analyst, and the spatial and socioeconomic characteristics of the area under investigation.

Valuing Ecosystem Services

Ecosystem services and the economic policies designed to ensure their conservation represent fundamental pillars of contemporary landscape analysis and environmental governance. These services underpin not only ecological balance but also the functioning of the national economy and the well-being of society as a whole. As awareness of their significance has grown, increasing attention has been directed toward the formulation of economic and policy instruments capable of safeguarding these services and promoting the sustainable use of natural resources.

Among these instruments, market-based mechanisms such as Payments for Ecosystem Services (PES) play a prominent role.

PES schemes provide direct (financial) incentives to landowners, farmers, or communities in exchange for conservation (or restoration) of ecosystems that supply vital services such as carbon sequestration, water purification, or soil fertility maintenance. In parallel, broader policy frameworks are emerging that explicitly integrate the socioecological value of ecosystem services into decision-making processes, ensuring that their contribution to both human and environmental well-being is adequately recognized and reflected in economic systems (Wegner, 2016).

This section examines how socioeconomic interventions and practical, place-based actions can support the protection, enhancement, and equitable management of ecosystem services, an objective now firmly embedded within international sustainability agendas. It is essential to emphasize that ecosystem services constitute the foundational infrastructure of economic activity, social development, and human welfare (Serra et al., 2018). Consequently, the creation of effective mechanisms to assign economic value to natural capital and to internalize ecological externalities represents an urgent and indispensable step toward ensuring the long-term sustainability of both ecosystems and societies.

Definitions and General Notions

The excessive exploitation of natural resources has become increasingly evident and, in many cases, irreversible. Contemporary environmental conditions are characterized by the growing frequency and intensity of biophysical processes that render the planet progressively less hospitable to human life. Within this context, the notion of "ecosystem services" provides a guiding framework for aligning human decisions and environmental policies with the broader goals of economic, social, and ecological sustainability. It underscores the diverse range of benefits that nature provides to humanity, and highlights the intricate interdependence between human activities and environmental systems.

One of the most authoritative definitions of ecosystem services was proposed by the Millennium Ecosystem Assessment, a global scientific initiative promoted by the United Nations to evaluate the state of the world's ecosystems and the services they deliver to society (Carpenter et al., 2006). According to this assessment, ecosystem services are defined as "the multiple benefits provided by ecosystems to humankind" (McMichael et al., 2005). Emerging in the academic literature during the 1980s, the concept gained prominence alongside growing awareness of anthropogenic impacts on the environment and living organisms. From this perspective, the relationship between humans and the environment is biunivocal, that is, characterized by mutual influence—emphasizing the need of establishing a balanced and sustainable regulatory framework.

Since ecosystems encompass both biotic and abiotic components, the terms *ecosystem services* and *environmental services* can, in certain contexts, be taken as equivalent. Within this framework, natural capital is defined as the complex system of living and non-living elements that collectively sustain the provision of ecosystem services that are temporally consistent and spatially diffuse. These services are indispensable not only for economic productivity but also for the overall enhancement of human well-being and quality of life. This assertion derives from the recognition that ecosystem components supply goods and services of vital importance to humanity such as clean air, pollination, and drinkable water arising from the dynamic interaction among various forms of (economic and natural) capital. Foremost among these is natural capital, which operates in conjunction with manufactured capital (e.g., buildings, infrastructure, machinery), human capital (e.g., knowledge, skills, and health), and social capital (e.g., networks, relationships, and institutions). Together, these forms of capital sustain the flow of ecosystem services that underpin both economic development and social welfare.

Nevertheless, it is essential to recognize that natural systems, while providing benefits, can also entail significant costs. This is particularly evident in the case of ecosystem restoration which, although aimed at rehabilitating degraded environments, can be extremely costly and, in certain circumstances, may induce irreversible alterations in landscape structure (Lugo, 2007). Such interventions may yield short-term gains—yet produce long-term drawbacks when assessed from broader temporal and ecological perspectives. Ecosystem services can be classified according to several methodological frameworks. One of the most widely adopted is that of Millennium Ecosystem Assessment, which identifies four major functional categories:

1. Provisioning services, including tangible goods such as food, clean water, fibers, and medicinal resources
2. Regulating services, encompassing processes such as climate regulation, air purification, and water-cycle maintenance
3. Cultural services, referring to non-material benefits such as spiritual, ethical, recreational, and aesthetic experiences, as well as contributions to social cohesion and identity
4. Supporting (or life-support) services, comprising essential ecological functions such as soil formation, nutrient cycling, and primary biomass production

A more recent and detailed framework is provided by the Common International Classification of Ecosystem Services (CICES), which organizes ecosystem services (Haines-Young & Potschin, 2012) into three macro-classes, namely "Provisioning," "Regulation and Maintenance," and "Cultural." These categories are further subdivided into divisions, groups, classes, and types. For instance, within the macro-class "Provisioning," the division "Biomass" includes the group "Cultivated Plants," the class "Cultivation of Plants for Nutrition," and the type "Cereals." Such classification systems are of fundamental importance for ecosystem accounting, as they provide operational tools for quantifying the contributions of ecosystems to society in both biophysical and monetary terms. They facilitate the integration of ecological data into decision-making processes, economic planning, and sustainability policies, thereby linking environmental functions with socioeconomic outcomes (Turkelboom et al., 2013). In addition, these frameworks support cost–benefit analysis, environmental accounting, and the enhancement of public awareness regarding the essential role of natural systems in supporting human welfare.

The protection and sustainable management of ecosystems are therefore crucial not only for the well-being of present generations, but also for safeguarding opportunities for future ones. Achieving these objectives requires collaboration among multiple stakeholders, including governments, local communities, businesses, land managers, and non-governmental organizations (NGOs). Only through coordinated and participatory governance can the complex challenges of ecosystem conservation and sustainable resource use be effectively addressed. It is equally important to recognize the interconnected nature of ecosystem services. Consequently, any policy (or intervention) designed to protect a specific service should take account of its interactions with others. For instance, forest conservation simultaneously contributes to climate regulation, timber production, and biodiversity preservation. Acknowledging such interdependencies is essential for designing integrated environmental policies capable of maintaining ecosystem balance, resilience, and multifunctionality over time.

Information Sources

Data sources constitute a reasonable and representative set of information from which data required for a specific (quantitative) analysis can be obtained. Quantitative data supporting the evaluation of ecosystem services may be derived from a wide range of sources, several of which have already been discussed. These include official documents, surveys, statistical datasets, interviews, public registers, government databases, technological sensor detections (e.g., satellite observations), and even social media platforms that capture spatial and behavioral information relevant to environmental research. Naturally, these sources differ substantially in both nature and purpose. They can be classified into different categories, namely (a) primary sources, which consist of data collected directly by the researcher or the organization conducting the study (e.g., raw datasets, interviews, questionnaires, and field observations), and (b) secondary sources, which include data previously gathered and published by other entities, such as scientific articles, research reports, corporate documents (balance sheets or technical reports), and institutional publications.

Further distinctions can be made according to additional characteristics such as whether the data are qualitative or quantitative, public or private, and internal or external to the organization conducting the analysis. In all cases, it is crucial to select data sources that are both reliable and relevant to the objectives of the study. This selection process should be guided by explicit methodological criteria and strategic considerations, supported by tools capable of integrating heterogeneous data originating from multiple sources. Increasingly, researchers should also account for technological and regulatory developments, including innovations in data storage and processing systems ("Big Data"), as well as cybersecurity protocols and privacy regulations governing data protection and the right to personal information.

Within the context of landscape and ecosystem analysis, several technological and institutional tools are particularly valuable. National and regional geoportals, together with various geolocation services, provide systematic access to satellite imagery and detailed mapping tools essential for spatial and temporal analysis. In addition, remote sensing technologies including the use of satellites, aerial photography, and drones, enable precise monitoring and characterization of both natural and agricultural ecosystems. Complementary to these sources, historical databases and land registry records supply longitudinal information on the evolution of landscapes, land-use dynamics, and property ownership patterns. When integrated, these datasets facilitate a comprehensive and multiscalar understanding of environmental change, forming a robust empirical foundation for the evaluation and management of ecosystem services.

The Economic Role of Ecosystem Services

Considering the benefits that the environment provides to humanity entirely free of charge, ecosystem services encompass a wide range of operational dimensions that societies often take for granted in everyday life. Yet, these services constitute the very foundation of human existence. They include the provision of essential resources such as food, drinking water, and raw materials as well as critical regulatory and supporting functions, including climate regulation, disease control, oxygen production, and flood mitigation, among others. Consequently, the role of ecosystem services is fundamental to human well-being, both in the present and for future generations, and to the sustainability of the planet. Human populations depend on ecosystems to meet their most essential needs. These needs encompass security understood as personal safety, equitable access to natural resources, and protection from environmental hazards—health, social cohesion (including mutual respect, cooperation, and solidarity) and, ultimately, freedom of choice and action, defined as the ability to pursue one's own goals and exercise self-determination within a stable and supportive environment (Ganivet, 2020).

In this sense, ecosystem services not only sustain biological and material life but also shape the social, cultural, and psychological dimensions of human well-being. By maintaining the delicate equilibrium between natural systems and human societies, they serve as the indispensable bridge linking ecological integrity with economic prosperity and social development. The conservation and sustainable use of ecosystem services are therefore essential to ensuring the long-term viability of natural resources and the overall health of environments. Such practices not only safeguard ecological integrity but also guarantee the continuity, reliability, and resilience of services that ecosystems provide to humanity (Roche & Campagne, 2017).

Moreover, ecosystem services play a role in protecting biodiversity and maintaining the stability of natural systems over time. Another significant aspect concerns the economic dimension of ecosystem services. These services underpin the functioning of the global economy by supporting productive sectors and market systems. For instance, the pollination of plants by insects constitutes an indispensable ecosystem service for agricultural production, as many crop species depend directly on this process for fertilization. Beyond pollination, ecosystems provide additional vital services such as pest control, soil fertility, and nutrient cycling that are fundamental to sustaining agricultural productivity. The absence (or degradation) of these services would inevitably lead to a substantial increase in production costs, with serious implications for food security at both local and global scales.

Similarly, ecosystem services generate significant economic benefits in the tourism sector (Campos et al., 2021), particularly in natural and protected areas. Ecosystems such as forests, coral reefs, wetlands, and marine reserves attract millions of visitors each year, thereby generating considerable economic returns through accommodation, guided tours, recreational activities, and associated services. In parallel, aquatic ecosystems provide essential habitats for marine organisms, forming the foundation of fishing and aquaculture industries. The degradation (or loss) of these habitats would therefore have direct (and severe) repercussions on the economies of coastal communities, as well as on local and regional food security.

Ecosystems play a fundamental role in the provision of drinking water through a variety of natural processes, including filtration, aquifer recharge, and the regulation of watercourses. These ecological functions ensure the purification and continuous renewal of freshwater resources, thereby sustaining both human populations and economic systems. However, the sustainability of these functions is closely tied to the responsible use and management of water resources. When overexploited or mismanaged, the ecosystems that perform these regulatory roles are subject to progressive degradation, ultimately jeopardizing the quality and availability of water (Salvati & Zitti, 2008).

The economic dimension of ecosystem services is therefore multidimensional and of critical importance, as various economic sectors depend either directly or indirectly on the availability of natural resources and the continuous provision of ecological functions (Gomes et al., 2021). For instance, agriculture, industry, and energy production all rely on healthy ecosystems to ensure a stable supply of clean water and raw materials. Conversely, the intrinsic degradation of these natural systems increases the costs of production, infrastructure maintenance, and public health protection. From this perspective, conserving ecosystems and maintaining their functional integrity can yield substantial long-term savings (Das, 2013). The integration of ecosystem services into economic planning and policy development thus represents an essential strategy for promoting social and economic sustainability over medium and long terms. Moreover, investing in the conservation and restoration of ecosystems generates tangible economic benefits that extend beyond immediate environmental gains (e.g., Fares et al., 2017). Restored or well-managed ecosystems can enhance agricultural and fisheries productivity, increase the attractiveness of tourism destinations, and strengthen resilience to natural disasters, thereby reducing economic vulnerability and promoting adaptive capacity (Mejjad & Rovere, 2021).

An equally important consideration concerns the potential for employment creation and the promotion of sustainable business models associated with ecosystem protection. The notion of "green jobs" refers to employment opportunities that both depend upon and contribute to the conservation and sustainable use of natural resources. For instance, sustainable forest management policies can stimulate the certified timber sector, while investments in renewable energy and ecosystem restoration projects can foster the emergence of a skilled and environmentally responsible workforce. In this sense, the preservation of ecosystems transcends environmental stewardship; it instead represents a strategic pathway for economic diversification, innovation, and social well-being (Salvati et al., 2012).

Preserving Ecosystem Services

Economic policies aimed at preserving ecosystem services are fundamental to ensuring the intrinsic sustainability of natural resources and the well-being of local and global communities. A diverse range of economic instruments has been developed to achieve these objectives, including the taxation of environmentally harmful activities, subsidies for conservation-oriented practices, Payments for Ecosystem Services (PES) schemes, and the creation of markets for ecosystem services. Each of these mechanisms contributes, in distinct yet complementary ways, to the integration of ecological values within economic systems (Ureta et al., 2022).

The taxation of environmentally damaging activities serves as a central regulatory tool for discouraging pollution and resource degradation. By increasing the costs associated with harmful actions, environmental taxes incentivize both producers and consumers to adopt more sustainable production and consumption behaviors. Among the most well-known instruments are Pigovian taxes, which are based on the principle that polluters should bear the costs of the environmental damage they cause (Banzhaf, 2020).

Such taxes may be proportional to the quantity (or intensity) of emissions, or to the estimated economic cost of the resulting environmental harm.

In addition to emission taxes, other categories include resource exploitation taxes, which internalize the environmental costs of extraction, processing, and waste management, and production taxes, applied to goods or services with negative environmental externalities. In contrast, grants and subsidies function as positive economic incentives designed to encourage conservation and the sustainable use of natural resources. These instruments can promote biodiversity protection, habitat restoration, sustainable agricultural practices, and climate change mitigation.

Grounded in the "polluter pays" principle, PES schemes offer financial compensation to landowners, farmers, or local communities in exchange for maintaining, restoring, or enhancing the ecosystem services their land provides such as carbon sequestration, watershed protection, or biodiversity conservation. These payments can be financed by public agencies, private corporations, or civil society organizations, and serve to internalize externalities by ensuring that the economic value of ecosystem services is explicitly recognized and rewarded. In the European context, the "polluter pays" principle represents a cornerstone of environmental regulation, designed to correct market failures by ensuring that the full cost of production reflects both private and environmental costs. Moreover, the emergence of local and regional markets for ecosystem services represents an innovative and increasingly significant component of environmental economic policy. Such markets generate new economic opportunities, particularly in rural and resource-dependent communities, while simultaneously increasing the perceived and market value of ecological functions (Thompson et al., 2017).

A prominent example is the carbon market, which enables the trading of carbon credits to incentivize emission reductions in accordance with international climate targets. Similarly, markets for tradable pollution permits establish legally defined emission quotas that can be exchanged among firms, thereby encouraging cost-effective pollution control and enhancing environmental accountability (Danley & Widmark, 2016). Ultimately, economic policies for ecosystem service conservation seek to embed natural capital within the broader framework of development planning. By aligning market mechanisms with ecological sustainability, these policies not only safeguard ecosystems for future generations, but also promote a redefinition of economic growth as rooted in basic principles such as resilience, equity, and long-term environmental balance.

Concluding Remarks

By supporting projects that align with long-term environmental and social objectives, subsidies help cover the initial costs of sustainable practices or reward actions that generate positive ecological externalities. Whether provided by governmental institutions, non-governmental organizations, or international funding mechanisms, such financial incentives strengthen environmental stewardship and facilitate the transition toward low-carbon and resource-efficient economies. By linking complexity theory with the study of socioecological systems, our work reveals the inherently non-linear and interdependent nature of human–environment interactions, stressing the necessity of integrated policies and educational strategies that foster sustainability at both local and global levels. Ultimately, we argue that a structural transition toward innovative economic paradigms (including circular economy) is essential to reconcile economic growth with ecological limits. Such a transformation represents a fundamental step toward establishing a more balanced, regenerative, and enduring relationship between society, the economy, and the environment.

References

Abdullah, S., Markandya, A., & Nunes, P. A. (2011). Introduction to economic valuation methods. In *Research tools in natural resource and environmental economics* (pp. 143–187). Springer.

Akhtar, N., Syakir Ishak, M. I., Bhawani, S. A., & Umar, K. (2021). Various natural and anthropogenic factors responsible for water quality degradation: A review. *Water, 13*(19), 2660.

AlSaggaf, A., & Jrade, A. (2023). ArcSPAT: An integrated building information modeling (BIM) and geographic information system (GIS) model for site layout planning. *International Journal of Construction Management, 23*(3), 505–527.

Bansal, V. K. (2020). Use of GIS to consider spatial aspects in construction planning process. *International Journal of Construction Management, 20*(3), 207–222.

Banzhaf, H. S. (2020). *A history of pricing pollution (or, why Pigouvian taxes are not necessarily Pigouvian) (no. w27683)*. National Bureau of Economic Research.

Baymuminova, N., Shermukhammedova, G., & Choi, J. G. (2023). Estimating the economic value of Ichan Kala using the contingent valuation method (CVM). *Sustainability, 15*(3), 2631.

Blanco, E., Lopez, M. C., & Walker, J. M. (2016). The opportunity costs of conservation with deterministic and probabilistic degradation externalities. *Environmental and Resource Economics, 64*(2), 255–273.

Boente, C., Gerassis, S., Albuquerque, M. T. D., Taboada, J., & Gallego, J. R. (2020). Local versus regional soil screening levels to identify potentially polluted areas. *Mathematical Geosciences, 52*(3), 381–396.

Borrego, A., & Skutsch, M. (2014). Estimating the opportunity costs of activities that cause degradation in tropical dry forest: Implications for REDD+. *Ecological Economics, 101*, 1–9.

Brookes, S. (2007). Walking with Anglo-Saxons: Landscapes of the dead in early Anglo-Saxon Kent. *Anglo-Saxon Studies in Archaeology and History, 14*, 143.

Campos, F. S., David, J., Lourenço-de-Moraes, R., Rodrigues, P., Silva, B., da Silva, C. V., & Cabral, P. (2021). The economic and ecological benefits of saving ecosystems to protect services. *Journal of Cleaner Production, 311*, 127551.

Cárdenas, O. B., Herrera Arango, J., Guerrero Lovera, C., & Helo Molina, E. (2022). Bridging research and practice to influence national policy: Afro-Colombians' territorial rights, from stagnation to implementation. *Bulletin of Latin American Research, 41*(3), 387–403.

Carpenter, S. R., DeFries, R., Dietz, T., Mooney, H. A., Polasky, S., Reid, W. V., & Scholes, R. J. (2006). Millennium ecosystem assessment: Research needs. *Science, 314*(5797), 257–258.

Carson, R. T., Flores, N. E., & Mitchell, R. C. (1999). The theory and measurement of passive-use value. In *Valuing environmental preferences: Theory and practice of the contingent valuation method in the US, EU, and developing countries* (pp. 97–130). Springer.

Churski, P., Herodowicz, T., Konecka-Szydłowska, B., & Perdał, R. (2021). *European regional development: Contemporary regional and local perspectives of socio-economic and socio-political changes*. Springer Nature.

Clough, P. W., & Meister, A. D. (1991). Allowing for multiple-site visitors in travel cost analysis. *Journal of Environmental Management, 32*(2), 115–125.

Cox, L. A., Jr. (2020). Implications of nonlinearity, confounding, and interactions for estimating exposure concentration–response functions in quantitative risk analysis. *Environmental Research, 187*, 109638.

Danley, B., & Widmark, C. (2016). Evaluating conceptual definitions of ecosystem services and their implications. *Ecological Economics, 126*, 132–138.

Das, S. (2013). *Travel cost method for environmental valuation*. Center of Excellence in Environmental Economics, Madras School of Economics, Dissemination Paper No. 23.

Doğan, T. G., & Eroğlu, E. (2021). Using GIS and the diversity indices: A combined approach to woody plant diversity in the urban landscape. *Vegetation Index and Dynamics, 9*, 1–10.

Ekbote, N., Dhanshetti, P., & Sakhrekar, S. (2023). Techniques of exploratory data analysis. *Madhya Pradesh Journal of Social Sciences, 28*(2), 5–12.

Elhashash, M., Albanwan, H., & Qin, R. (2022). A review of mobile mapping systems: From sensors to applications. *Sensors, 22*(11), 4262.

Engström, E., & Kolk, M. (2024). Projecting environmental impacts with varying population, affluence and technology using IPAT: Climate change and land use scenarios. *Vienna Yearbook of Population Research, 2024*, 1–29.

Eshet, T., Ayalon, O., & Shechter, M. (2006). An inclusive comparative review of valuation methods for assessing environmental goods and externalities. *International Journal of Business Environment, 1*(2), 190–210.

Fares, S., Bajocco, S., Salvati, L., Camarretta, N., Dupuy, J.-L., Xanthopoulos, G., Guijarro, M., Madrigal, J., Hernando, C., & Corona, P. (2017). Characterizing potential wildland fire fuel in live vegetation in the Mediterranean region. *Annals of Forest Science, 74*, 1–14.
Ferrara, A., Kosmas, C., Salvati, L., Padula, A., Mancino, G., & Nolè, A. (2020). Updating the MEDALUS-ESA framework for worldwide land degradation and desertification assessment. *Land Degradation & Development, 31*(12), 1593–1607.
Ferretti, M., Marchetto, A., Arisci, S., Bussotti, F., Calderisi, M., Carnicelli, S., Cecchini, G., Fabbio, G., Bertini, G., Matteucci, G., de Cinti, B., Salvati, L., & Pompei, E. (2014). On the tracks of nitrogen deposition effects on temperate forests at their southern European range–an observational study from Italy. *Global Change Biology, 20*(11), 3423–3438.
Fradelos, E. C., Papathanasiou, I. V., Mitsi, D., Tsaras, K., Kleisiaris, C. F., & Kourkouta, L. (2014). Health based geographic information systems (GIS) and their applications. *Acta Informatica Medica, 22*(6), 402.
Ganivet, E. (2020). Growth in human population and consumption both need to be addressed to reach an ecologically sustainable future. *Environment, Development and Sustainability, 22*(6), 4979–4998.
Gomes, E., Inácio, M., Bogdzevič, K., Kalinauskas, M., Karnauskaitė, D., & Pereira, P. (2021). Future land-use changes and its impacts on terrestrial ecosystem services: A review. *Science of the Total Environment, 781*, 146716.
Haines-Young, R., & Potschin, M. (2012). Common international classification of ecosystem services (CICES, version 4.1). *European Environment Agency, 33*, 107.
Halbac-Cotoara-Zamfir, R., Smiraglia, D., Quaranta, G., Salvia, R., Salvati, L., & Giménez-Morera, A. (2020). Land degradation and mitigation policies in the Mediterranean region: A brief commentary. *Sustainability, 12*(20), 8313.
Halls, A. J. E. (1997). Wetlands, biodiversity and the Ramsar convention: The role of the convention on wetlands in the conservation and wise use of biodiversity. In *Ramsar convention bureau* (Vol. 13). Gland.
Han, C., Tang, F., Ma, T., Gu, L., & Tong, Z. (2022). Construction quality evaluation of asphalt pavement based on BIM and GIS. *Automation in Construction, 141*, 104398.
Hasti, F., Mamkhezri, J., McFerrin, R., & Pezhooli, N. (2023). Optimal solar photovoltaic site selection using geographic information system–based modeling techniques and assessing environmental and economic impacts: The case of Kurdistan. *Solar Energy, 262*, 111807.
Hidano, N. (2002). The economic valuation of the environment and public policy: A hedonic approach. In *The economic valuation of the environment and public policy*. Edward Elgar.
Howari, F. M., & Ghrefat, H. (2021). Geographic information system: Spatial data structures, models, and case studies. In *Pollution assessment for sustainable practices in applied sciences and engineering* (pp. 165–198). Butterworth-Heinemann.
Isacs, L., Håkansson, C., Lindahl, T., & Gunnarsson-Östling, U. (2021). 'I didn't count WTP as part of the value': Respondents' perspective on willingness to pay—An inductive approach. *Ecological Economics, 183*, 106950.
Isah, A. M., Iliyasu, I., & Barde, U. (2023). Spatial distribution and analysis of secondary schools in Damaturu Metropolis. *Fane-Fane International Multidisciplinary Journal, 7*(2), 444–457.
Jackson, S., Finn, M., & Scheepers, K. (2014). The use of replacement cost method to assess and manage the impacts of water resource development on Australian indigenous customary economies. *Journal of Environmental Management, 135*, 100–109.
Krishnendu, J., & Patra, B. (2025). Environmental impacts of economic growth: A STIRPAT analysis using machine learning algorithms. *Sustainable Futures, 9*, 100404.
Kroes, E. P., & Sheldon, R. J. (1988). Stated preference methods: An introduction. *Journal of Transport Economics and Policy*, 11–25.
Lake, I. R., Lovett, A. A., Bateman, I. J., & Day, B. (2000). Using GIS and large-scale digital data to implement hedonic pricing studies. *International Journal of Geographical Information Science, 14*(6), 521–541.

Lesser, V. M., Rawlings, J. O., Spruill, S. E., & Somerville, M. C. (1990). Ozone effects on agricultural crops: Statistical methodologies and estimated dose–response relationships. *Crop Science, 30*(1), 148–155.

Lester, J. N., Cho, Y., & Lochmiller, C. R. (2020). Learning to do qualitative data analysis: A starting point. *Human Resource Development Review, 19*(1), 94–106.

Li, X., Li, S., Li, C., Shi, J., & Wang, N. (2023). The impact of high-quality development on ecological footprint: An empirical research based on STIRPAT model. *Ecological Indicators, 154*, 110881.

Lohwasser, J., Schaffer, A., & Brieden, A. (2020). The role of demographic and economic drivers on the environment in traditional and standardized STIRPAT analysis. *Ecological Economics, 178*, 106811.

Lü, G., Batty, M., Strobl, J., Lin, H., Zhu, A. X., & Chen, M. (2019). Reflections and speculations on the progress in geographic information systems (GIS): A geographic perspective. *International Journal of Geographical Information Science, 33*(2), 346–367.

Lugo, E. (2007). Ecosystem services, the millennium ecosystem assessment, and the conceptual difference between benefits provided by ecosystems and benefits provided by people. *Journal of Land Use & Environmental Law, 23*, 243.

Luvai, A., Obiero, J., & Omuto, C. (2022). Soil loss assessment using the revised universal soil loss equation (RUSLE) model. *Applied and Environmental Soil Science, 2022*(1), 2122554.

Maghsoudi, A., Mansouri, S. A., & Haghir, S. (2024). Territorial landscape: Explaining the relationship between the concept of territory and landscape. *MANZAR: The Scientific Journal of Landscape, 16*(68), 48–59.

Matthews, G. V. T. (1993). *The Ramsar convention on wetlands: Its history and development*. Ramsar Convention Bureau.

McMichael, A., Scholes, R., Hefny, M., Pereira, E., Palm, C., & Foale, S. (2005). Linking ecosystem services and human well-being. *Ecosystems and Human Well-Being, 5*, 43–60.

Mejjad, N., & Rovere, M. (2021). Understanding the impacts of blue economy growth on deep-sea ecosystem services. *Sustainability, 13*(22), 12478.

Moffett, D. B., Mumtaz, M. M., Sullivan, D. W., Jr., & Whittaker, M. H. (2022). General considerations of dose–effect and dose–response relationships. In *Handbook on the toxicology of metals* (pp. 299–317). Academic Press.

Mogorovich, P., & Salvatori, E. (2022). Historical GIS. In *Handbook of digital public history* (p. 419). Springer Nature.

Mounsey, S., Waqa, G., McKenzie, B., Reeve, E., Webster, J., Bell, C., & Thow, A. M. (2022). Strengthening implementation of diet-related non-communicable disease prevention strategies in Fiji: A qualitative policy landscape analysis. *Globalization and Health, 18*(1), 79.

Munafo, M., Norero, C., Sabbi, A., & Salvati, L. (2010). Soil sealing in the growing city: A survey in Rome, Italy. *Scottish Geographical Journal, 126*(3), 153–161.

Mundher, R., Abu Bakar, S., Maulan, S., Mohd Yusof, M. J., Al-Sharaa, A., Aziz, A., & Gao, H. (2022). Aesthetic quality assessment of landscapes as a model for urban forest areas: A systematic literature review. *Forests, 13*(7), 991.

Nisa, R. U., Tantray, A. Y., Kouser, N., Allie, K. A., Wani, S. M., Alamri, S. A., & Shah, A. A. (2021). Influence of ecological and edaphic factors on biodiversity of soil nematodes. *Saudi Journal of Biological Sciences, 28*(5), 3049–3059.

Niu, Q., Yu, L., Jie, Q., & Li, X. (2020). An urban eco-environmental sensitive areas assessment method based on variable weights combination. *Environment, Development and Sustainability, 22*, 2069–2085.

Noor, N. M., Asmawi, M. Z., & Abdullah, A. (2015). Sustainable urban regeneration: GIS and hedonic pricing method in determining the value of green space in housing area. *Procedia—Social and Behavioral Sciences, 170*, 669–679.

Osborne, S. A., Mills, G., Hayes, F., Ainsworth, E. A., Büker, P., & Emberson, L. (2016). Has the sensitivity of soybean cultivars to ozone pollution increased with time? An analysis of published dose–response data. *Global Change Biology, 22*(9), 3097–3111.

Plottu, E., & Plottu, B. (2007). The concept of Total economic value of environment: A reconsideration within a hierarchical rationality. *Ecological Economics, 61*(1), 52–61.

Poorter, H., Niinemets, Ü., Walter, A., Fiorani, F., & Schurr, U. (2010). A method to construct dose–response curves for a wide range of environmental factors and plant traits by means of a meta-analysis of phenotypic data. *Journal of Experimental Botany, 61*(8), 2043–2055.

Pouta, E., Rekola, M., Kuuluvainen, J., Li, C. Z., & Tahvonen, O. (2002). Willingness to pay in different policy-planning methods: Insights into respondents' decision-making processes. *Ecological Economics, 40*(2), 295–311.

Roche, P. K., & Campagne, C. S. (2017). From ecosystem integrity to ecosystem condition: A continuity of concepts supporting different aspects of ecosystem sustainability. *Current Opinion in Environmental Sustainability, 29*, 63–68.

Ruankham, W. (2025). Air pollution and housing market valuation: A spatial hedonic pricing approach to welfare loss estimation. *International Journal of Housing Markets and Analysis*, Vol. ahead-of-print No. ahead-of-print. https://doi.org/10.1108/IJHMA-03-2025-0054

Salvati, L., & Zitti, M. (2008). Regional convergence of environmental variables: Empirical evidences from land degradation. *Ecological Economics, 68*(1–2, 162), –168.

Salvati, L., Perini, L., Sabbi, A., & Bajocco, S. (2012). Climate aridity and land use changes: A regional-scale analysis. *Geographical Research, 50*(2), 193–203.

Sannigrahi, S., Zhang, Q., Joshi, P. K., Sutton, P. C., Keesstra, S., Roy, P. S., & Sen, S. (2020). Examining effects of climate change and land use dynamic on biophysical and economic values of ecosystem services of a natural reserve region. *Journal of Cleaner Production, 257*, 120424.

Serra, P., Saurí, D., & Salvati, L. (2018). Peri-urban agriculture in Barcelona: Outlining landscape dynamics vis-à-vis socio-environmental functions. *Landscape Research, 43*(5), 613–631.

Shores, J., Daniel, B., & Faircloth, W. B. (2024). The experience of inspiration in natural landscapes: Awe, wonder, sublimity, and Bergson's qualitative multiplicity. *The Journal of Experimental Education, 47*(2), 256–274.

Sillmann, J., Aunan, K., Emberson, L., Büker, P., Van Oort, B., O'Neill, C., & Brisebois, A. (2021). Combined impacts of climate and air pollution on human health and agricultural productivity. *Environmental Research Letters, 16*(9), 093004.

Simoniello, T., Coluzzi, R., D'Emilio, M., Imbrenda, V., Salvati, L., Sinisi, R., & Summa, V. (2022). Going conservative or conventional? Investigating farm management strategies between economic and environmental sustainability in southern Italy. *Agronomy, 12*(3), Article 597.

Smiraglia, D., Tombolini, I., Canfora, L., Bajocco, S., Perini, L., & Salvati, L. (2019). The latent relationship between soil vulnerability to degradation and land fragmentation: A statistical analysis of landscape metrics in Italy, 1960–2010. *Environmental Management, 64*(2), 154–165.

Thakuriah, G. (2023). GIS-based revised universal soil loss equation for estimating annual soil erosion: A case of lower Kulsi basin, India. *SN Applied Sciences, 5*(3), 81.

Thompson, B. S., Primavera, J. H., & Friess, D. A. (2017). Governance and implementation challenges for mangrove forest payments for ecosystem services (PES): Empirical evidence from The Philippines. *Ecosystem Services, 23*, 146–155.

Tombolini, I., Munafò, M., & Salvati, L. (2016). Soil sealing footprint as an indicator of dispersed urban growth: A multivariate statistical approach. *Urban Research & Practice, 9*(1), 1–15.

Turkelboom, F., Raquez, P., Dufrêne, M., Raes, L., Simoens, I., Jacobs, S., & Keune, H. (2013). CICES going local: Ecosystem services classification adapted for a highly populated country. In *Ecosystem services* (pp. 223–247). Elsevier.

Ureta, J. C., Motallebi, M., Vassalos, M., Seagle, S., & Baldwin, R. (2022). Estimating residents' WTP for ecosystem services improvement in a payments for ecosystem services (PES) program: A choice experiment approach. *Ecological Economics, 201*, 107561.

Vardopoulos, I., Ioannides, S., Georgiou, M., Voukkali, I., Salvati, L., & Doukas, Y. E. (2023). Shaping sustainable cities: A long-term GIS-emanated spatial analysis of settlement growth and planning in a coastal Mediterranean European city. *Sustainability, 15*(14), Article 11202.

Vinci, S., Vardopoulos, I., & Salvati, L. (2023). A tale of a shrinking city? Exploring the complex interplay of socio-demographic dynamics in the recent development of Attica, Greece. *Cities, 132*, Article 104089.

Walker, M., & Mondello, M. J. (2007). Moving beyond economic impact: A closer look at the contingent valuation method. *International Journal of Sport Finance, 2*(3), 149–160.

Wani, Z. A., Bhat, J. A., Negi, V. S., Satish, K. V., Siddiqui, S., & Pant, S. (2022). Conservation priority index of species, communities, and habitats for biodiversity conservation and their management planning: A case study in Gulmarg Wildlife Sanctuary, Kashmir Himalaya. *Frontiers in Forests and Global Change, 5*, 995427.

Wegner, G. I. (2016). Payments for ecosystem services (PES): A flexible, participatory, and integrated approach for improved conservation and equity outcomes. *Environment, Development and Sustainability, 18*(3), 617–644.

Westhoek, H., Van Zeijts, H., Witmer, M., Van den Berg, M., Overmars, K., Van der Esch, S., & Van der Bilt, W. (2012). *Greening the CAP: An analysis of the effects of the European Commission's proposals for the Common Agricultural Policy 2014–2020*. European Commission.

Wyatt, P. (2009). Replacement cost and market value. *Journal of Property Investment & Finance, 27*(6), 593–602.

Xu, X., Dong, R., Li, Z., Jiang, Y., & Genovese, P. V. (2024). Research on visual experience evaluation of fortress heritage landscape by integrating SBE–SD method and eye movement analysis. *Heritage Science, 12*(1), 281.

Zambon, I., & Salvati, L. (2019). Metropolitan growth, urban cycles, and housing in a Mediterranean country, 1910s–2010s. *Cities, 95*, Article 102412.

Zambon, I., Serra, P., Grigoriadis, E., Carlucci, M., & Salvati, L. (2017). Emerging urban centrality: An entropy-based indicator of polycentric development and economic growth. *Land Use Policy, 68*, 365–371.

Zambon, I., Ferrara, A., Salvia, R., Mosconi, E. M., Fici, L., Turco, R., & Salvati, L. (2018). Rural districts between urbanization and land abandonment: Undermining long-term changes in Mediterranean landscapes. *Sustainability, 10*(4), Article 1159.

Zhang, Z., Xu, Z., & Cheng, G. (2003). The updated development and application of contingent valuation method (CVM). *Advances in Earth Science, 18*(3), 454–463.

Zhang, F., Wang, X. H., Nunes, P. A., & Ma, C. (2015). The recreational value of Gold Coast beaches, Australia: An application of the travel cost method. *Ecosystem Services, 11*, 106–114.

Zhang, Z., Mu, X., & Hu, G. (2024). Analysis of influencing factors of energy consumption in Beijing: Based on the IPAT model. *Environment, Development and Sustainability, 26*(5), 12569–12588.

Zlatanova, S., & Stoter, J. (2006). The role of DBMS in the new generation GIS architecture. In *Frontiers of geographic information technology* (pp. 155–180). Springer.

MIX
Papier aus verantwortungsvollen Quellen
Paper from responsible sources
FSC® C105338

If you have any concerns about our products,
you can contact us on
ProductSafety@springernature.com

In case Publisher is established outside the EU,
the EU authorized representative is:
Springer Nature Customer Service Center GmbH
Europaplatz 3, 69115 Heidelberg, Germany

Printed by Libri Plureos GmbH
in Hamburg, Germany